SpringerBriefs in Applied Sciences and Technology

Thermal Engineering and Applied Science

Series Editor

Francis A. Kulacki, Department of Mechanical Engineering, University of Minnesota, Minneapolis, MN, USA

More information about this series at http://www.springer.com/series/8884

Sujoy Kumar Saha • Hrishiraj Ranjan
Madhu Sruthi Emani • Anand Kumar Bharti

Introduction to Enhanced Heat Transfer

 Springer

Sujoy Kumar Saha
Mechanical Engineering Department
Indian Institute of Engineering,
Science & Technology, Shibpur
Howrah, West Bengal, India

Hrishiraj Ranjan
Mechanical Engineering Department
Indian Institute of Engineering,
Science & Technology, Shibpur
Howrah, West Bengal, India

Madhu Sruthi Emani
Mechanical Engineering Department
Indian Institute of Engineering,
Science & Technology, Shibpur
Howrah, West Bengal, India

Anand Kumar Bharti
Mechanical Engineering Department
Indian Institute of Engineering,
Science & Technology, Shibpur
Howrah, West Bengal, India

ISSN 2191-530X   ISSN 2191-5318   (electronic)
SpringerBriefs in Applied Sciences and Technology
ISSN 2193-2530   ISSN 2193-2549   (electronic)
SpringerBriefs in Thermal Engineering and Applied Science
ISBN 978-3-030-20742-7   ISBN 978-3-030-20740-3   (eBook)
https://doi.org/10.1007/978-3-030-20740-3

This Springer imprint is published by the registered company Springer Nature Switzerland AG
The registered company address is: Gewerbestrasse 11, 6330 Cham, Switzerland

Contents

Nomenclature

A Area
A_{ni} Nominal surface area based on fin root diameter
d Diameter
d_{i} Fin root diameter
d_{o} Outer diameter
D Diameter
e Fin height
E Electric field strength
f Friction factor
f_{e} EHD force density
g Gravitational acceleration
G Mass velocity
h Convective heat transfer coefficient
h_{lv} Latent heat
K Thermal conductivity
L Characteristics dimension; length
L Length
M_{d} Masuda number
N Number of channels
n_{s} Number of fins
Nu Nusselt number
p Pressure
p_{f} Fin pitch normal to the fins
P Pressure
ΔP Pressure drop
q Heat flux
Q Heat transfer rate
Re Reynolds number
Re_{1} Liquid Reynolds number
t Time

T	Temperature
U	Mean velocity
V	Volume
x	Vapor quality
X	Effective quality

Greek Symbols

α	Heat transfer coefficient; apex angle of the fin
β	Helix angle
δ	Liquid film thickness
Δp	Pressure drop
ΔT	Temperature difference
ε	Permittivity
μ	Dynamic viscosity
ρ	Density
σ	Surface tension

Subscripts

ave	Average
ev	Evaporation
in	Inlet
l	Liquid
s	Saturated
sub	Subcooled
v	Vapor

Abbreviation

CHF	Critical heat flux
CNT	Carbon nanotube
EHD	Electrohydrodynamic
ONB	Onset of nucleate boiling

Additional Nomenclature

a_1–a_4	Constants
A	Constant
A_{ch}	Channel cross-sectional area, m^2

A_h	Heated inside area, m^2
B	Constants in empirical correlations
c_p	Specific heat capacity at constant pressure (J/kg K)
$C\ C_0$	Parameter in empirical correlations
Ca	Capillary number
Co	Confinement number
$D\ C_1–C_2$	Constants
D_e	Equivalent diameter, same as hydraulic diameter, m
D_h	Hydraulic diameter, m
E	Bubble diameter, channel diameter, or tube diameter, m
E_1	Parameter
E_2	Parameter
F	Force, N
F^0	Force per unit length, N/m
F_S	Surface tension force
g	Gravitational acceleration, m/s^2
G	Mass flux, kg/m^2 s
h	Heat transfer coefficient, W/m^2 K
h_{LV}	Latent heat of vaporization, J/kg
Δh_{Sub}	Subcooling enthalpy, J/kg
j	Superficial velocity, m/s
k	Thermal conductivity, W/m K
K	Inlet subcooling parameter in empirical correlations
$K_1–K_3$	Parameters in empirical correlations
K_2	Nondimensional group
L	Length, m
$\underline{m}$	Mass flow rate, kg/s
n	Parameter in empirical correlation
P_L	Liquid pressure, Pa
P_V	Vapor pressure, Pa
q	Heat flux, W/m^2
q_{CHF}	Heat flux at CHF, W/m^2
q_{CHF0}	CHF based on channel heated inside area for zero inlet subcooling
q_{CHF01} to q_{CHF04}	Parameters in empirical correlations
qv	Heat flux, W/m^2
Q	Volumetric flow rate, m^3/s
r	Radii, m
Re	Reynolds number
Re_{crit}	Critical Reynolds number
S	Slip ratio
T	Temperature, $^\circ C$
ΔT	Temperature difference, $^\circ C$
U	Velocity, m/s

We	Weber number
x	Quality (defined as the ratio of the mass flow rate of vapor to the total mass flow rate)
y	Parameter
Y_{shah}	Shah's correlation parameter

Greek Symbols

α	Void fraction, or half angle at the corner included by two channel walls
α_{Hom}	Homogeneous void fraction
β	Volume flow fraction
δ	Film thickness
δ_0	Initial film thickness
δ_t	Thermal boundary layer thickness, m
o	Void fraction, same as α
θ	Contact angle, °
θ_1	Contact angle on one channel surface, °
θ_2	Contact angle on the adjoining surface, °
θ_R	Receding contact angle, °
λ	Parameter in empirical correlations
μ	Dynamic viscosity, kg/m s
ρ	Density, kg/m^3
ρ_m	Average density
o	Surface tension, N/m

Subscripts

c	Cavity
CHF	At critical heat flux condition
crit	Critical
eq	Equivalent
exit	At the exit section
G	Gaseous phase
I	Inertia
inlet	At the inlet section
L	Liquid phase
M	Evaporation momentum
max	Maximum
min	Minimum
ONB	Onset of nucleate boiling
r	Receding
S	Surface tension

Sat	Saturation
Sub	Subcooling
τ	Shear (viscous)
V	Vapor
wall	Channel wall

Chapter 1
Heat Transfer Fundamentals for Design of Heat Transfer Enhancement Devices

Scientists and researchers have keen interest in heat-exchanging devices. The reason behind this is that it has broad applications in daily life and industries. Air conditioner, shell and tube heat exchangers, boilers, condensers, radiators, heaters, furnaces, cooling towers, solar collectors, heating, ventilation and air conditioning (HVAC) equipment and refrigerators use heat-exchanging devices. The automotive industries, chemical industries, petrochemical industries and refrigeration industries are the prime beneficiaries of heat exchangers.

Every industry, small or large, has heat transfer processes for various operations. The use of heat transfer enhancement techniques makes the heat exchangers compact and efficient, but at the same time cost increases proportionally. The viable adaptation of heat exchangers is the hotspot for last two decades. The researchers revise the techniques time to time for better performance and to obtain optimised designs.

Bergles et al. (1983) proposed 13 techniques for enhancement of heat transfer. Some basic techniques are surface coatings, extended surfaces like fins, rough surfaces, coiled tubes, square and helical ribs, electrostatic fields, mechanical aids and surface vibration. The enhancement techniques are broadly divided into two groups: active technique and passive technique. These will be discussed later in this chapter.

Roughness techniques are one of the several enhancement techniques. This enhancement of heat transfer techniques for a single-phase flow results in mixing of the boundary layer rather than increase in heat transfer surface area. The contribution of mixing of boundary layer is more significant than that of increased surface area. The basic idea behind the use of coating surfaces is to promote condensation on the surface. Extended surfaces are commonly applied in heat exchangers. Benefit of using extended surfaces over plain surface is that it provides higher heat transfer coefficient than that in case of the plain surfaces, for the same surface area.

Typically, short fins are used for liquids and longer fins are used for gases because liquids have higher heat transfer coefficient than gases in general. Some models of extended surfaces are presented in Fig. 1.1 and it shows the geometry and terminology used for the channels with plate-fins. All the above techniques are used for

© The Author(s), under exclusive license to Springer Nature Switzerland AG 2020
S. K. Saha et al., *Introduction to Enhanced Heat Transfer*, SpringerBriefs in Applied Sciences and Technology, https://doi.org/10.1007/978-3-030-20740-3_1

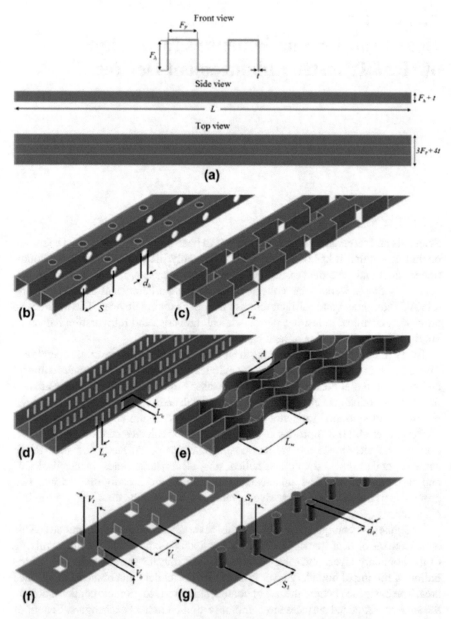

Fig. 1.1 Plate-fin channels with different geometrical channels (Webb and Kim 2005)

disturbing the boundary layers near wall. Thus, practical use of devices like tapes, ribs and insert coils disturbs the boundary layer and enhances heat transfer. This is the fundamental approach used for the heat transfer enhancement. Nowadays researchers use a combination of different techniques for optimisation and compactness of heat-exchanging devices.

Fig. 1.2 Integral spiral corrugation in a circular duct (Pal and Saha 2014)

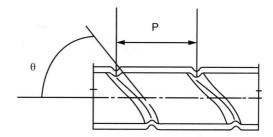

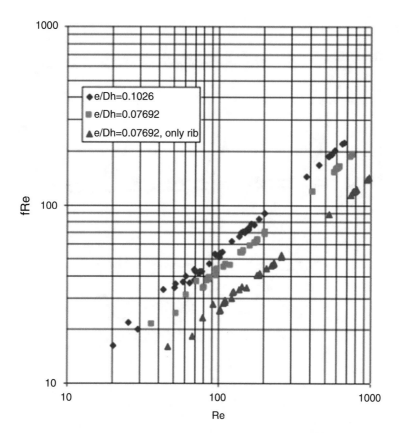

Fig. 1.3 Effect of rib height, $y = 2.5$, thl $= 0.1053$, $h = 60$, $P/e = 35.6481$ (friction factor) (Pal and Saha 2015)

Pal and Saha (2014) (Fig. 1.2), Pal and Saha (2015) (Fig. 1.3), Saha et al. (2012) (Fig. 1.4), Saha (2010) (Fig. 1.5), Saha (2012) (Fig. 1.6), Darzi et al. (2013) (Fig. 1.7), Naik et al. (2014) (Fig. 1.8), Murugesan et al. (2011) (Figs. 1.9 and 1.10), Nanan et al. (2014) (Fig. 1.11), Promvonge (2008) (Fig. 1.12), Promvonge and Eiamsa-ard (2006), Promvonge et al. (2012) (Fig. 1.13) and Thianpong et al. (2009) (Fig. 1.14) worked on combined techniques and got improved results.

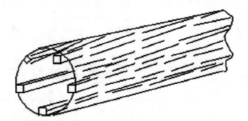

Fig. 1.4 Axially ribbed circular duct (Saha et al. 2012)

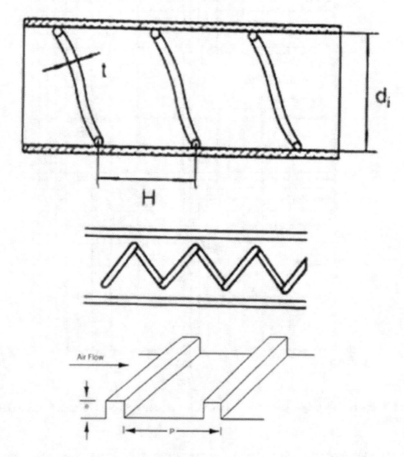

Fig. 1.5 Wire coil inserts: wires touching tube wall, wires displaced from tube wall and transverse ribs (Saha 2010)

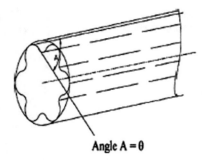

Angle A = θ

Fig. 1.7 Helically corrugated tube (Darzi et al. 2013)

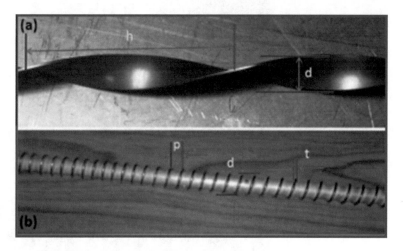

Fig. 1.8 Images of inserts: (**a**) twisted tape and (**b**) wire coil (Naik et al. 2014)

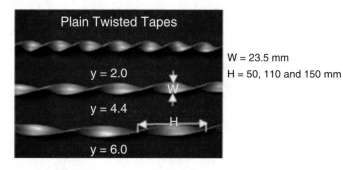

Fig. 1.9 Plain twisted tapes (Murugesan et al. 2011)

Fig. 1.10 Plain twisted tapes with V-cut (Murugesan et al. 2011)

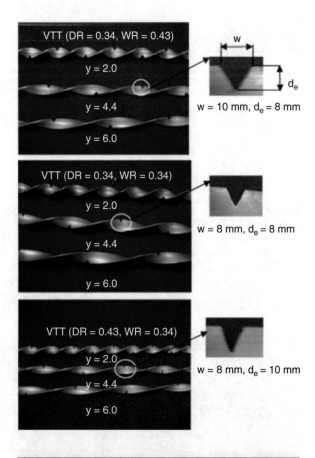

Fig. 1.11 Plain perforated tubes and helical perforated tubes (Nanan et al. 2014)

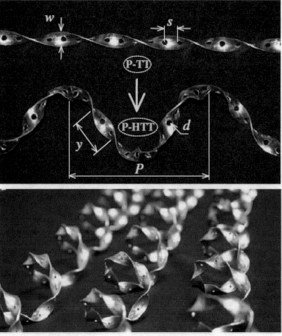

Fig. 1.12 Combination of
wire coil and twisted tape
(Promvonge 2008)

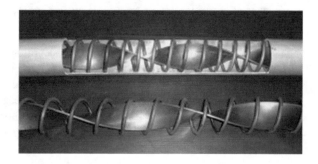

Fig. 1.13 Helical-ribbed
tube fitted with twin twisted
tapes (Promvonge et al.
2012)

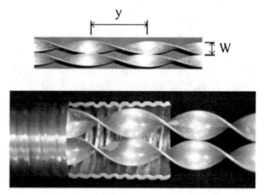

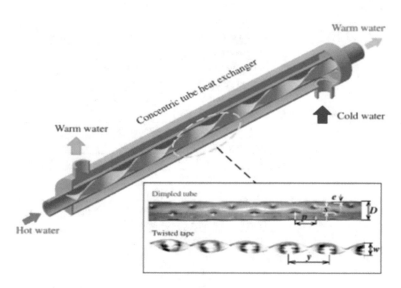

Fig. 1.14 Dimpled tube fitted with twisted tape (Thianpong et al. 2009)

1.1   Passive and Active Enhancement Techniques

Enhancement techniques are being upgraded according to technological developments. Bergles (2017) reported different techniques that have been used for the enhancement of convective heat transfer. He classified the heat transfer technology, used in industries, into three generations which have been presented in Table 1.1.

The first-generation heat transfer techniques commonly used bare tubes, the second generation progressed to plain fins, the third generation shifted to vortex generators on fins and extended surface and the fourth generation moved to compound enhancement techniques. The heat transfer enhancement techniques are divided into two groups, namely, active techniques and passive techniques. Compound enhancement techniques are summation of two or more individual active or passive techniques. The heat transfer performance using compound techniques improves tremendously compared to that for individual techniques.

A large amount of work on heat transfer enhancement techniques has been reported in the literature. Figure 1.15 shows the increase in number of citations on heat transfer enhancement till the year 2000. As the number of papers on heat and mass transfer enhancement has been rapidly increasing over years, Bergles and Manglik (2013) presented a detailed review covering both the single-phase and two-phase heat transfer enhancement phenomena. They reviewed the literature on active, passive and compound techniques.

The classification of various heat transfer augmentation techniques has been shown in Table 1.2. The use of rough surfaces for augmentation of single-phase

Table 1.1 Three generations of heat transfer technology (Bergles 2017)

Tube-and-plate fins, single phase	
First generation	Bare tube
Second generation	Plain fins
Third generation	Longitudinal vortex generators on fins
In-channel, single phase	
First generation	Smooth channel
Second generation	2-D roughness
Third generation	3-D roughness
Outside tubes, boiling	
First generation	Smooth tube
Second generation	2-D fins
Third generation	3-D fins and metallic matrices
In-tube, evaporation	
First generation	Smooth tubes
Second generation	Massive fins and inserts
Third generation	Micro-fins
Outside tubes, condensing	
First generation	Smooth tubes
Second generation	2-D fins
Third generation	3-D fins and metallic matrices

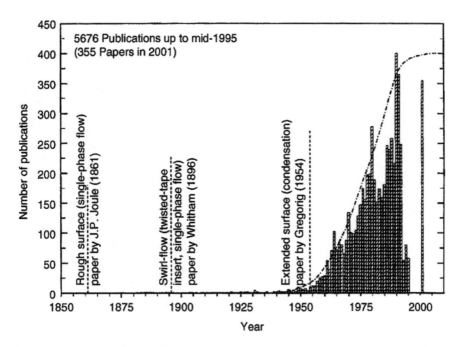

Fig. 1.15 Citations on heat transfer augmentation vs. year of publication. Illustrates status as of 2001 (Manglik and Bergles 2004)

Table 1.2 Classification of heat transfer augmentation techniques (Bergles and Manglik 2013)

Passive techniques	Active techniques
Treated surfaces	Mechanical aids
Rough surfaces	Surface vibration
Extended surfaces	Fluid vibration
Displaced enhancement devices	Electrostatic fields
Swirl-flow devices	Injection
Coiled tubes	Suction
Surface tension devices	Jet impingement
Additives for liquids	
Additives for gases	
Compound enhancement	

Two or more passive and/or active techniques used together
Wavy plate-fins with punched-tab vortex generators
Rotating internally finned tubes

heat transfer has been discussed by Webb et al. (1971, 1972), Burgess and Ligrani (2005), Lee et al. (2012) and Nishida et al. (2012). The rough surfaces are generally used to perturb the laminar sublayer present in turbulent flow regime. The extended surfaces are used to enhance heat transfer rates in heat exchangers by increasing the heat transfer surface area. Manglik and Bergles (2013), T'Joen et al. (2011), Xie and

Fig. 1.16 Different models
of louvered fins (T'Joen
et al. 2011)

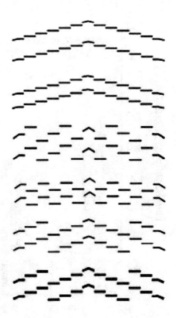

Sunden (2011), Wang et al. (2012), Manglik (2006) and Lindstedt and Karvinen (2012) studied extensively the heat transfer enhancement characteristics of extended surfaces.

Manglik and Jog (2009), Kuo and Peles (2009), Jones et al. (2009), Vasiliev et al. (2012), Xu et al. (2012), Drelich and Chibowski (2010), Rioboo et al. (2009), Takata et al. (2009) and Kananeh et al. (2010) worked with treated surfaces for heat transfer enhancement especially for boiling and condensation. From these works it has been observed that the effect of using wetting and nonwetting treatment on surfaces deteriorates with time. The six models of louvered fins have been shown in Fig. 1.16.

The louvered fins are popularly used for automobile engine cooling channels and heat exchangers. Akbari et al. (2009), Razani et al. (2001), Garrity et al. (2010), Hu et al. (2011) and Sachdeva et al. (2010) investigated the performance of displaced enhancement devices. Swirl-flow devices are one of the most extensively used techniques for enhancing both single-phase and two-phase heat transfer. They are used to disturb the entire flow region. Kanizawa and Ribatski (2012), Chen et al. (2009), Tarasevich et al. (2011), Bishara et al. (2009), Patel et al. (2012), Itaya et al. (2010) and Javed et al. (2010) reported the performance of various swirl-flow devices for heat transfer augmentation.

The swirl behaviour of flow through an axially twisted oval tube has been shown in Fig. 1.17. Neshumayev and Tiikma (2000, 2007), Tarasevich et al. (2011), Zimparov et al. (2012) and Chien and Hwang (2012) presented different models of compound inserts for heat transfer enhancement. Figure 1.18 shows different twisted-tape configurations with wire wraps.

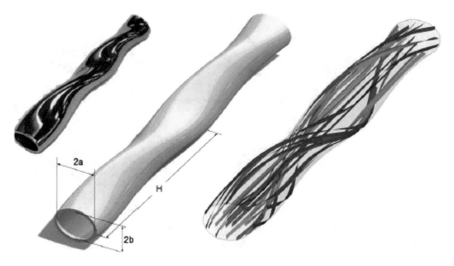

Fig. 1.17   Swirl flow through an axially twisted oval tube (Bishara et al. 2009)

1.2   Benefits of Enhancement

The heat transfer enhancement is required for heat-exchanging devices. For convective heat transfer mode, the basic heat transfer equation is given by

$$Q = hA\Delta T \tag{1.1}$$

where Q is the rate of heat transfer, h is the convective heat transfer coefficient and ΔT is the temperature difference. Heat transfer enhancement ratio, E_r, is the ratio of the (hA) of an enhanced surface to that of a plain surface or basic surface. Thus,

$$E_r = \frac{hA}{(hA)_p} \tag{1.2}$$

The rate of heat transfer for a constant temperature difference can be increased by two ways:

(a) Increasing "h" without changing physical area (A)
(b) Increasing heat transfer surface area (A) without changing "h"
(c) Increasing both "h" and "A"

   For the enhanced surface, the thermal resistance per unit tube length is reduced and this can be used for one of the three purposes: size reduction giving a smaller heat exchanger, increased conductance which will give reduced temperature difference and increased thermodynamic process efficiency and this finally gives a savings

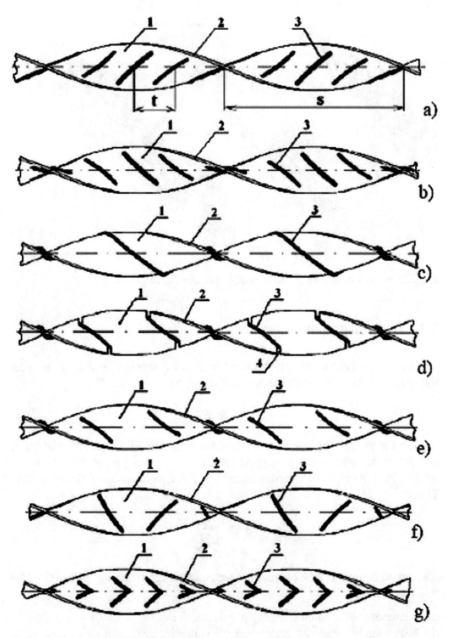

Fig. 1.18 Twisted tapes with wire wraps (Tarasevich et al. 2011)

of operating cost. Also, there will be increased heat exchange rate for fixed fluid inlet temperatures. Thirdly, pumping power can be reduced for fixed heat duty when the heat exchanger operates at a velocity smaller than the competing plain surface. This requires increased frontal area which is a negative effect.

The first case can be taken care of by surface roughness inside tube, duct, etc., whereas the second case can be achieved with finned tubes. The third case can be dealt with by using fins with coils, screw tapes, etc. These techniques result in heat transfer enhancement many times for the same pressure drop. In some cases, researchers have even found 300% increase in heat transfer rate over that in a plain tube.

However, it should be kept in mind that all geometries are not practical due to the difficulty in manufacturing complex geometries using required materials. The possible techniques have been discussed by Webb (1980). Table 1.3 shows these techniques with their relevance. Thus, enhancement techniques, either active or passive or compound techniques, reduce the operating cost by large amount. However, passive techniques are more significant over active techniques as active techniques are relatively noisy, their reliability is less, they are little costlier and they have safety issues.

Table 1.3 Active and passive enhancement techniques (Bergles and Webb, 1985)

	Single-phase natural convection	Single-phase forced convection	Pool boiling	Forced-convection boiling	Condensation	Mass transfer
Passive techniques (no external power required)						
Treated surfaces	–	–	149	17	53	+
Rough surfaces	7	418	62	65	65	29
Extended surface	23	416	75	53	175	33
Displaced enhancement devices	+	59	4	17	6	15
Swirl-flow devices	+	140	–	83	17	10
Coiled tubes	+	142	–	50	6	9
Surface tension devices	30	–	12	1	–	+
Additives for liquids	3	22	61	37	–	6
Additives for gases	+	211	–	–	5	0
Active techniques (external power required)						
Mechanical aids	16	60	30	7	23	18
Surface vibration	52	30	11	2	9	11
Fluid vibration	44	127	15	5	2	39
Electric or magnetic fields	50	53	37	10	22	22
Injection or suction	6	25	7	1	6	2
Jet impingement	–	17	2	1	–	2
Compound techniques	2	50	4	4	4	2

Note: – not applicable; + no citations located

References

Akbari MM, Murata A, Mochizuki S, Saito H, Iwamoto K (2009) Effects of vortex generator arrangements on heat transfer enhancement over a two-row fin-and-tube heat exchanger. J Enhanc Heat Transf 16(4):315–329

Bergles AE (2017) Endless frontier, or mature and routine of enhanced heat transfer. J Enhanc Heat Transf 24(1–6):329–338

Bergles AE, Manglik RM (2013) Current progress and new developments in enhanced heat and mass transfer. J Enhanc Heat Transf 20(1):1–15

Bergles AE, Nirmalan V, Junkhan GH, Webb RL (1983) Bibliography on augmentation of convective heat and mass transfer-II (No. ISU-ERI-AMES-84221). Iowa State Univ. of Science and Technology Ames (USA). Heat Transfer Lab

Bergles AE, Webb RL (1985) A guide to the literature on convective heat transfer augmentation, in advances in enhanced heat transfer - 1985, ASME, HTD Vol. 43, pp. 81–90

Bishara F, Jog MA, Manglik RM (2009) Heat transfer and pressure drop of periodically fully developed swirling laminar flows in twisted tubes with elliptical cross sections. InASME 2009 international mechanical engineering congress and exposition, Jan 1. American Society of Mechanical Engineers, pp 1131–1137

Burgess NK, Ligrani PM (2005) Effects of dimple depth on channel Nusselt numbers and friction factors. J Heat Transfer 127(8):839–847

Chen HB, Zhou Y, Yin J, Yan J, Ma Y, Wang L, Cao Y, Wang J, Pei J (2009) Single organic microtwist with tunable pitch. Langmuir 25(10):5459–5462

Chien LH, Hwang HL (2012) An experimental study of boiling heat transfer enhancement of mesh-on-fin tubes. J Enhanc Heat Transf 19(1):75–86

Darzi AAR, Farhadi M, Sedighi K, Aallahyari S, Delavar MA (2013) Turbulent heat transfer of Al_2O_3–water nanofluid inside helically corrugated tubes: numerical study. Int Commun Heat Mass Transfer 41:68–75

Drelich J, Chibowski E (2010) Superhydrophilic and superwetting surfaces: definition and mechanisms of control. Langmuir 26(24):18621–18623

Garrity PT, Klausner JF, Mei R (2010) Performance of aluminum and carbon foams for air side heat transfer augmentation. J Heat Transfer 132(12):121901

Hu WL, Zhang YH, Wang LB (2011) Numerical simulation on turbulent fluid flow and heat transfer enhancement of a tube bank fin heat exchanger with mounted vortex generators on the fins. J Enhanc Heat Transf 18(5):361–374

Itaya Y, Kobayashi N, Nakamiya T (2010) Okara drying by pneumatically swirling two-phase flow in entrained bed riser with enlarged zone. Dry Technol 28(8):972–980

Javed KH, Mahmud T, Purba E (2010) The CO_2 capture performance of a high-intensity vortex spray scrubber. Chem Eng J 162(2):448–456

Jones RJ, Pate DT, Thiagarajan N, Bhavnani SH (2009) Heat transfer and pressure drop characteristics in dielectric flow in surface-augmented microchannels. J Enhanc Heat Transf 16(3):225–236

Kananeh AB, Rausch MH, Leipertz A, Fröba AP (2010) Dropwise condensation heat transfer on plasma-ion-implanted small horizontal tube bundles. Heat Transfer Eng 31(10):821–828

Kanizawa FT, Ribatski G (2012) Two-phase flow patterns and pressure drop inside horizontal tubes containing twisted-tape inserts. Int J Multiphase Flow 47:50–65

Kuo CJ, Peles Y (2009) Flow boiling of coolant (HFE-7000) inside structured and plain wall microchannels. J Heat Transfer 131(12):121011

Lee YO, Ahn J, Kim J, Lee JS (2012) Effect of dimple arrangements on the turbulent heat transfer in a dimpled channel. J Enhanc Heat Transf 19(4):359–367

Lindstedt M, Karvinen R (2012) Optimization of isothermal plate fin arrays with laminar forced convection. J Enhanc Heat Transf 19(6):535–547

Manglik RM, Bergles AE (2004) Enhanced heat and mass transfer in the new millennium: A review of the 2001 literature. J Enhanc Heat Transf Vol. 11, pp. 87–118

Manglik RM (2006) On the advancements in boiling, two-phase flow heat transfer, and interfacial phenomena. J Heat Transfer 128(12):1237–1242

Manglik RM, Bergles AE (2013) Characterization of twisted-tape-induced helical swirl flows for enhancement of forced convective heat transfer in single-phase and two-phase flows. J Thermal Sci Eng Appl 5(2):021010

Manglik RM, Jog MA (2009) Molecular-to-large-scale heat transfer with multiphase interfaces: current status and new directions. J Heat Transfer 131(12):121001

Murugesan P, Mayilsamy K, Suresh S, Srinivasan PSS (2011) Heat transfer and pressure drop characteristics in a circular tube fitted with and without V-cut twisted tape insert. Int Commun Heat Mass Transfer 38(3):329–334

Naik MT, Fahad SS, Sundar LS, Singh MK (2014) Comparative study on thermal performance of twisted tape and wire coil inserts in turbulent flow using CuO/water nanofluid. Exp Therm Fluid Sci 57:65–76

Nanan K, Thianpong C, Promvonge P, Eiamsa-Ard S (2014) Investigation of heat transfer enhancement by perforated helical twisted-tapes. Int Commun Heat Mass Transfer 52:106–112

Neshumayev D, Tiikma T (2000) Radiation heat transfer of turbulator inserts in gas heated channels. Heat Transf Res 39(5):403–412

Neshumayev D, Tiikma T (2007) Review of compound passive heat transfer enhancement techniques. In: Proc. of the 5th Baltic heat transfer conf. on advances in heat transfer, Sept 19, pp 410–424

Nishida S, Murata A, Saito H, Iwamoto K (2012) Compensation of three-dimensional heat conduction inside wall in heat transfer measurement of dimpled surface by using transient technique. J Enhanc Heat Transf 19(4):331–341

Pal PK, Saha SK (2014) Experimental investigation of laminar flow of viscous oil through a circular tube having integral spiral corrugation roughness and fitted with twisted tapes with oblique teeth. Exp Therm Fluid Sci 57:301–309

Pal S, Saha SK (2015) Laminar fluid flow and heat transfer through a circular tube having spiral ribs and twisted tapes. Exp Therm Fluid Sci 60:173–181

Patel P, Manglik RM, Jog MA (2012) Swirl-enhanced laminar forced convection through axially twisted rectangular ducts–part 1. Fluid flow. J Enhanc Heat Transf 19(5):423–436

Promvonge P (2008) Thermal augmentation in circular tube with twisted tape and wire coil turbulators. Energy Convers Manag 49(11):2949–2955

Promvonge P, Eiamsa-Ard S (2006) Heat transfer enhancement in a tube with combined conical-nozzle inserts and swirl generator. Energy Convers Manag 47(18–19):2867–2882

Promvonge P, Pethkool S, Pimsarn M, Thianpong C (2012) Heat transfer augmentation in a helical-ribbed tube with double twisted tape inserts. Int Commun Heat Mass Transfer 39(7):953–959

Razani A, Paquette JW, Montoya B, Kim KJ (2001) A thermal model for calculation of heat transfer enhancement by porous metal inserts. J Enhanc Heat Transf 8(6):411–420

Rioboo R, Marengo M, Dall'Olio S, Voué M, De Coninck J (2009) An innovative method to control the incipient flow boiling through grafted surfaces with chemical patterns. Langmuir 25 (11):6005–6009

Sachdeva G, Kasana KS, Vasudevan R (2010) Heat transfer enhancement by using a rectangular wing vortex generator on the triangular shaped fins of a plate-fin heat exchanger. Heat Transfer Asian Res 39(3):151–165

Saha SK (2010) Thermohydraulics of laminar flow through rectangular and square ducts with axial corrugation roughness and twisted tapes with oblique teeth. J Heat Transfer 132:081701

Saha SK (2012) Thermohydraulics of laminar flow of viscous oil through a circular tube having axial corrugations and fitted with centre-cleared twisted-tape. Exp Therm Fluid Sci 38:201–209

Saha SK, Swain BN, Dayanidhi B (2012) Friction and thermal characteristics of laminar flow of viscous oil through a circular tube having axial corrugations and fitted with helical screw-tape inserts. J Fluids Eng 134:051210

T'Joen C, Huisseune H, Willockx A, Caniere H, De Paepe M (2011) Combined experimental and numerical flow field study of inclined louvered fins. Heat Transfer Eng 32(2):176–188

Takata Y, Hidaka S, Kohno M (2009) Wettability improvement by plasma irradiation and its applications to phase-change phenomena. Heat Transfer Eng 30(7):549–555

Tarasevich SE, Yakovlev AB, Giniyatullin AA, Shishkin AV (2011) Heat and mass transfer in tubes with various twisted tape inserts. In: ASME 2011 international mechanical engineering congress and exposition, Jan 1. American Society of Mechanical Engineers, pp 697–702

Thianpong C, Eiamsa-Ard P, Wongcharee K, Eiamsa-Ard S (2009) Compound heat transfer enhancement of a dimpled tube with a twisted tape swirl generator. Int Commun Heat Mass Transfer 36(7):698–704

Vasiliev LL, Zhuravlyov AS, Shapovalov A (2012) Heat transfer enhancement in mini channels with micro/nano particles deposited on a heat-loaded wall. J Enhanc Heat Transf 19(1):13–24

Wang FX, Shi GQ, Xu T, Zhang ZG (2012) Experimental investigation on condensation heat transfer of R410A on single horizontal petal-shaped finned tube. J Enhanc Heat Transf 19 (6):527–533

Webb RL (1980) Special surface geometries for heat transfer augmentation. Applied Science Publishers, England, pp 179–215

Webb RL, Eckert ER, Goldstein R (1971) Heat transfer and friction in tubes with repeated-rib roughness. Int J Heat Mass Transfer 14(4):601–617

Webb RL, Eckert ER, Goldstein RJ (1972) Generalized heat transfer and friction correlations for tubes with repeated-rib roughness. Int J Heat Mass Transfer 15(1):180–184

Webb RL, Kim NY (2005) Principles of enhanced heat transfer. Taylor and Francis, NY

Xie G, Sunden B (2011) Conjugated analysis of heat transfer enhancement of an internal blade tip-wall with pin-fin arrays. J Enhanc Heat Transf 18(2):149–165

Xu ZG, Qu Z, Zhao CY, Tao WQ (2012) Experimental study of pool boiling heat transfer on metallic foam surface with U-shaped and V-shaped grooves. J Enhanc Heat Transf 19 (6):549–559

Zimparov V, Petkov VM, Bergles AE (2012) Performance characteristics of deep corrugated tubes with twisted-tape inserts. J Enhanc Heat Transf 19(1):1–11

Chapter 2
Active and Passive Techniques: Their Applications

2.1 Active Techniques

Active techniques require external power sources like electric field, acoustic field and surface vibrations for heat transfer enhancement. Also, mechanical aids used for stirring the fluids by any mechanical means come under active techniques. In industries, apparatus integrated with rotating heat exchanger ducts are usually used for heat transfer augmentation. Surface vibration technique with different frequency ranges has been used for enhancing heat transfer rate. The vibration propels the small droplets onto the heated surface where thin-film evaporation occurs. The impingement of small droplets promotes "spray cooling". Electrostatic fields are applied for greater bulk mixing of dielectric fluid. Jet impingements, bubble injection, fluid vibration and application of magnetic field are some other active techniques.

Tada et al. (1997) theoretically and experimentally studied the characteristics of electrohydrodynamic (EHD) forces for heat transfer enhancement. They accounted for the wire electrode position, primary velocity and applied voltage for results. They discussed the potential application of ionic wind and visualised the flow pattern with a series of wire electrodes installed parallel as well as perpendicular to primary flow direction. Marco and Velkoff (1963) worked on EHD natural convective heat transfer. Ohadi et al. (2000), Laohalertdecha et al. (2007) and Webb and Kim (2005) presented a general review on EHD-enhanced heat transfer.

Yonggang et al. (2006) studied to enhance the heat transfer rate by using ionic wind on a heated vertical plate having needle-to-plate configuration. Sahebi and Alemrajabi (2014) adopted electrohydrodynamic active enhancement technique. They investigated experimentally the effect of electric field on an inclined heated plate by using convective heat transfer mode.

Figure 2.1 shows that by increasing the applied voltage, EHD heat transfer coefficient increases. Baxi and Ramachandran (1969) experimentally investigated the free and forced convective heat transfer caused by vibration of sphere made up of

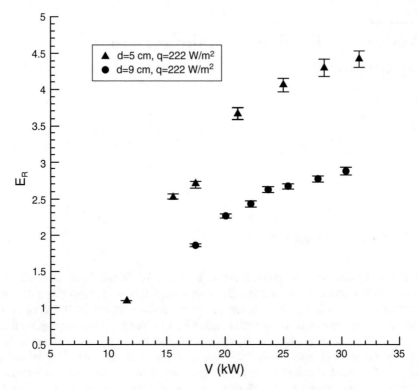

Fig. 2.1 Effect of applied voltage on enhancement ratio (Sahebi and Alemrajabi 2014)

copper. They concluded that, for free convection, the effect of vibration on Nusselt number was significantly high up to seven times the value obtained by free convection without vibration.

Martin (1977) and Goldstein and Timmers (1982) compared single-jet-impingement heat transfer performance with that of jet arrays. Webb and Ma (1995), Lee and Lee (2000), Stevens and Webb (1991, 1992), Elison and Webb (1994), Garimella and Nenaydykh (1996), Garimella and Rice (1995), Lee et al. (2004), Lou et al. (2005), O'Donovan and Murray (2007) and Brignoni and Garimella (2000) investigated single circular impingement jets and gave information regarding heat transfer from fluid such as air, FC-77 and water.

Uddin et al. (2019) reported computational results of heat transfer augmentation using turbulent jet impingement on a flat surface. They compared three methods for assessing the heat transfer rates. The methods were swirl flow generation by guide vanes, forced vibration having some frequency and amplitude and vortex shedding occurrence due to placing of cylinder in flow direction. They performed large eddy simulations and concluded that inserts in the jet nozzle (a passive excitation) give better heat transfer than others. The active excitation and swirl generation method showed much lower heat transfer rates than the passive excitation method.

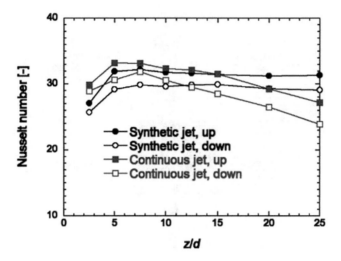

Fig 2.2 Comparison of jets (Fukiba et al. 2018)

Fukiba et al. (2018) worked on heat transfer enhancement of heated cylinder conjugated with synthetic jet impingement from multiple orifices and compared own results with those of others. They compared flow characteristics with piccolo tube and heated cylinder and calculated Nusselt number for heated cylinder with time and observed 2–3 times higher heat transfer rates than those without jets. They also concluded that the performance of upward directional jet was good and profound effectiveness of synthetic jets was obtained when they were placed at greater distances than that of continuous jets. Figure 2.2 shows the variation of average Nusselt number with the distance between orifice and cylinder for synthetic jets and continuous jets which favour the results.

Kumar and Pattamatta (2015) worked on the numerical simulation of convective heat transfer enhancement with three different configurations of jet impingement cooling porous block. The detailed physical configurations, coordinate system and three different positions of porous obstacle with jet impingement cooling system are shown in Fig. 2.3.

Numerical studies are conducted for three different configurations in jet impingement with a porous block as a heat sink in the stagnation zone, as a flow obstacle along the bottom plate and as a flow obstacle along the top plate away from the stagnation zone. Simulated results show that the convective heat transfer can only be enhanced in the case of flow obstacles along the top confining wall away from the stagnation zone. However, it was not clear whether the enhancement was either due to porous heat sink obstacles alone or due to variation of geometric parameters used for comparison.

Metzger et al. (1984), Metzger and Sahm (1986), Fan and Metzger (1987), Rao et al. (2004), Rao and Prabhu (2004), Luo and Razinsky (2009) and Chen et al. (2011) studied the effect of guide vanes on pressure loss and heat transfer in the bend

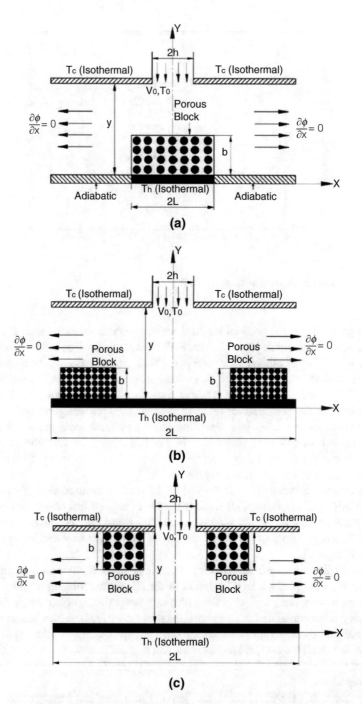

Fig. 2.3 Jet impingement with porous media (**a**) as porous heat sinks, (**b**) as flow obstacles along the bottom plate and (**c**) as flow obstacles along the top wall (Kumar and Pattamatta 2015)

channel. Schabacker et al. (1998) and Son et al. (2002) found that jet impingement in bend regions increased the velocity, which is the main factor to enhance the heat transfer.

Ökten and Biyikoglu (2018) studied the enhancement of heat transfer using bubble injection, which is one of the active heat transfer augmentation techniques. They investigated the heat transfer enhancement in a thermal storage tank with air bubble injection. They reported that the enhancement in overall heat transfer between the storage fluid and the environment had increased five times and that between the storage fluid and the heat transfer fluid had increased two times.

Kitagawa et al. (2010), Dizaji and Jafarmadar (2014), Funfschilling and Li (2006), Nouri and Sarreshtehdari (2009) and González-Altozano et al. (2015) studied the impact of bubble injection on heat transfer augmentation. Warrier and Dhir (2013) studied the heat removal from electronic chips using miniature channels, sprays and microjets. They provided a detailed review with performance comparison of the three methods in enhancing heat transfer from chips.

2.2 Passive Techniques

Passive techniques do not require any external source of energy like electric field and mechanical aids. The heat transfer rate can be augmented by surface coatings, rough surfaces, extended surfaces, surface tension devices, inserts, coiled tubes, and square and helical ribs. It can also be increased by using additives for liquids as well as for gases.

Rabas et al. (1994) used three-dimensional helically dimpled tubes and calculated the influence of roughness shape and spacing on tubes. They observed that greater roughness curvature and closer roughness spacing increased the performance significantly. Raineri et al. (1996), Kang et al. (1997), Rainieri and Pagliarini (1997), Saha (2010), Saha et al. (2012) and Pal and Saha (2014) worked on laminar flow regime. Other works on dimpled tubes have been carried out by Vulchanov et al. (1991), Zimparov et al. (1991), Nozu et al. (1995), Macbain et al. (1997), Lee et al. (1998), Salim et al. (1999), Wang et al. (2000), Chen et al. (2001) and Laohalertdecha and Wongwises (2010).

Rainieri et al. (1996) experimentally studied laminar internal flow of a Newtonian fluid under the uniform heat wall flux condition using internal helical ridging. They investigated the heat transfer and pressure drop in spiral stainless steel tubes and concluded that heat transfer rate for cross-corrugated tubes was much higher than that of smooth tube. They also found that the transition from laminar to the turbulent flow occurred at very low Reynolds number around 2000.

Saha and Saha (2013) experimented with circular channel having integral helical rib roughness and circular channel fitted with helical screw tapes. They evaluated friction factor and Nusselt number for laminar flow of viscous oil through the channels. They found that a combination of helical screw tape and helical rib roughness enhanced the heat transfer rate substantially over the individual

Fig. 2.4 Schematic of
circular tube with wire-coil
insert (Rout and Saha 2013)

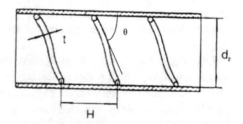

techniques. Also, they concluded that friction factor increased with increase in rib
helix angle.

Rout and Saha (2013) worked together experimentally on a circular tube
having wire coils and helical screw-tape inserts in the laminar flow regime. They
used log regression and developed predictive friction factor and Nusselt number
correlations. They found that a combination of wire coils with helical screw-tape
inserts performed significantly better than that of individual techniques. Figures 2.4
and 2.5 show the wire-coil inserts and helical screw tape and twisted tape,
respectively.

Khoshvaght-Aiabadi et al. (2016) used wavy plate-fins (WPFs), commonly used
in plate-fin exchanger for performance improvement and size reduction. They
compared three passive techniques, namely, perforations, winglets and nanofluids.
The parameters and flow specifications were kept fixed. They concluded that the
winglets performed well because of efficient mixing of fluid with respect to perfo-
ration and nanofluids. Figures 2.6 and 2.7 show comparison between perforated
winglets, winged wavy plate-fins, 0.1% nanofluids and 0.3% nanofluids.

Mohammed et al. (2019) numerically investigated and studied the thermal and
hydraulic characteristics of two-phase forced convection using nanofluids. They
simulated nanofluid flow with four nanoparticles (Al_2O_3, CuO, SiO_2 and ZnO) in
base fluid (water) with convergent and divergent conical ring inserts. They concluded that SiO_2 nanoparticles performed best among four. They claimed that
performance enhancement by divergent ring inserts is 365% and two-phase mixture
model is more precise. Figure 2.8 shows the SiO_2 performance and Fig. 2.9 shows
the plot in favour of divergent ring inserts in two-phase flow.

Saha and Paul (2018) simulated Al_2O_3-water and TiO_2-water nanofluids to
investigate the flow characteristics. They used single-phase model and constant
heat flux boundary conditions. They found that size of nanoparticles and their
concentrations and Reynolds number govern the heat transfer rate as well as entropy
generation. They proposed new correlations to calculate average Nusselt number.

Kumar et al. (2016, 2017), Patil (2017), Qu et al. (2017), Naik and Vinod (2018),
Gupta et al. (2018), Chang and Yu (2018), Kareem et al. (2015), Noh and Suh
(2014), Leontiev et al. (2016) and Abraham and Vedula (2016) also used different
passive techniques for heat transfer augmentation.

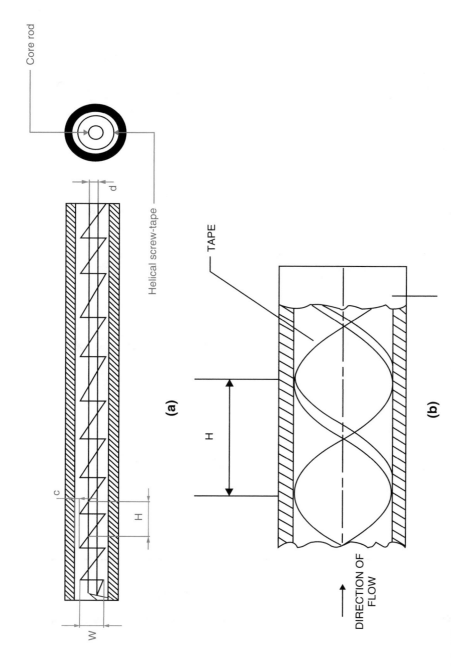

Fig. 2.5 Schematic representation of (**a**) helical screw tape and (**b**) twisted tape (Rout and Saha 2013)

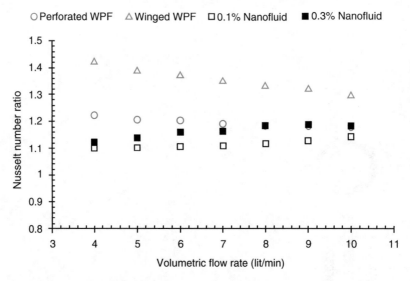

Fig 2.6 Performance evaluation of passive techniques, Nusselt number vs. volumetric flow rate (Khoshvaght-Aiabadi et al. 2016)

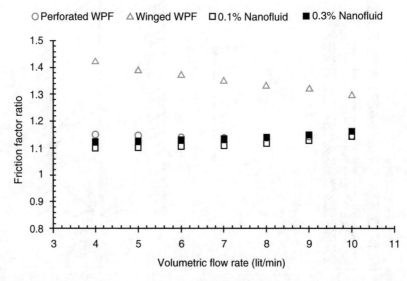

Fig. 2.7 Performance evaluation of passive techniques, friction factor vs. volumetric flow rate (Khoshvaght-Aliabadi et al. 2016)

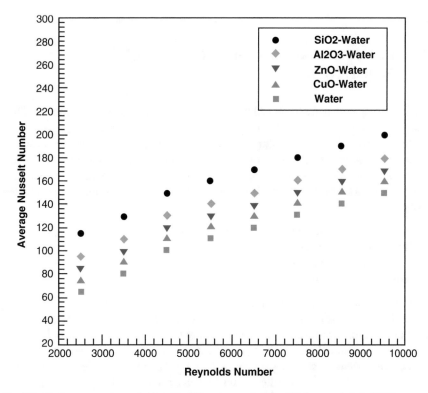

Fig. 2.8 Performance of nanofluids with different nanoparticles (Mohammed et al. 2019)

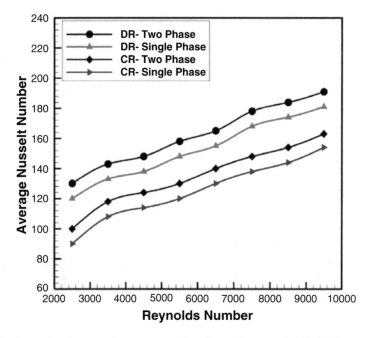

Fig. 2.9 Comparison between single phase and two phase (Mohammed et al. 2019)

2.3   Commercial Applications of Heat Transfer Augmentation

The heat transfer enhancement techniques have wide commercial applications. They include heat exchangers; cooling towers; refrigeration units; heating, ventilation and air conditioning; distillation columns; etc. The heat and mass transfer occurs simultaneously in some devices like cooling towers and distillation columns. Commercially, the heat transfer enhancement techniques are used for heat exchangers and this book focuses on the same. The heat exchangers are classified into four main groups:

- Shell-and-tube heat exchangers or tube banks
- Fin-and-tube heat exchangers
- Plate-fin heat exchangers
- Plate-type heat exchangers

The heat transfer enhancement in heat exchangers can be achieved by enhancing basic flow geometries, namely, internal flow in circular section, external flow through or across the tubes, flow in noncircular section (stacked plate-fin type) and flow between parallel plates (plate-type heat exchangers). Clear understanding of various enhancement special geometries may be made from Table 2.1. The commercially exploited and used heat transfer enhancement techniques are classified below.

Table 2.1 Different enhancement techniques and their applications (Webb and Kim 2005)

	Avail.	Mode			Materials	Performance potential
		CV	B	C		
Inside tubes						
Metal coatings	Yes	5	2	5	Al, Cu, St	Hi
Integral fins	Yes	2	1	1	Al, Cu	Hi
Fluxes	Yes	4	4	1	Al, Ga	Mod
Integral	Yes	1	2	2	Cu, St	Hi
Wine coil	Yes	1	2	2	Any	Mod
Displaced promoters	Yes	2	2	2	Any	Mod (Lam)
Twisted tape	Yes	1	2	2	Any	Mod
Outside circular tubes						
Coatings						
Metal	Yes	5	1	3	Al, Cu, St	Hi (B)
Nonmetal	No	5	4	4	"Te"	Mod
Roughness (integral)	Yes	3	2	2	Al, Cu	Hi (B)
Roughness (attached)	Yes	2	3	3	Any	Mod (CV)
Axial fins	Yes	1	5	5	Al, St	Hi (CV)

(continued)

Table 2.1   (continued)

		Mode				
	Avail.	CV	B	C	Materials	Performance potential
Transverse fins						
Gases	Yes	1	5	5	Al, Cu, St	Hi
Liquids, two-phase	Yes	1	1	1	Any	Hi
Flutes						
Integral	Yes	5	5	1	Al, Cu	Hi
Non-integral	Yes	5	5	3	Any	Hi
Plate-fin heat exchanger						
Metal coatings	Yes	5	3	5	Al	Hi (B)
Surface roughness	Yes	4	4	4	Al	Hi (B)
Configured channel	Yes	1	1	1	Al, St	Hi
Integrated fins	Yes	1	1	1	Al, St	
Flutes	No	5	5	4	Al	Mod
Plate-type heat exchanger						
Metal coatings	No	5	4	4	St	Hi (B)
Surface roughness	No	4	4	4	St	Mod (CV)
Configured channel	Yes	1	3	3	St	Hi (CV)
Use code					*Heat transfer mode*	
1. Common use					CV convection	
2. Limited use					Boiling	
3. Scene special cases					Condensation	
4. Essentially no use						
5. Non-applicable						

2.3.1   Corrugated Tubes, Dimpled Tubes and Three-Dimensional Roughness

Figure 2.10 shows the corrugated tube. The figure is of Wolverine Korodense tube, typically used in industries. O'Brien and Sparrow (1982), Asako and Faghri (1987), Yang et al. (1997), Rush et al. (1999), Nishimura et al. (1986, 1990), Zhang et al. (2004), Motamed Ekitesabi et al. (1987), Metwally and Manglik (2004) and Vyas et al. (2010) worked on various corrugated tube configurations.

Isaev et al. (2018) numerically studied heat transfer augmentation in a dimpled narrow tube. The enhancement is attributed to the vortex generated due to dimples. They compared the performance of oval-trench dimpled tube with spherical dimpled tube. They reported that the tube with oval-trench dimples showed better thermal performance than the tube with spherical dimples.

Rabas et al. (2017) presented the efficiency index for dimples on heated surface. Different geometry of the dimples and different spacing between them have been considered. The efficiency index can be defined as the ratio of enhancement of heat transfer and increase in friction factor. They reported that dimples when placed closely and having larger roughness curvature resulted in greater enhancement.

Fig. 2.10 Corrugated tube
(Webb and Kim 2005)

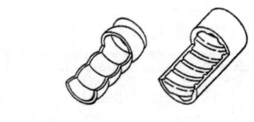

Fig. 2.11 Double-wall
cooling structure (Luo et al.
2015)

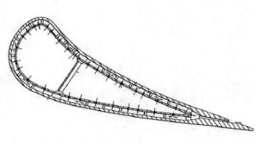

Fig 2.12 Lamilloy cooling
structures (Luo et al. 2015)

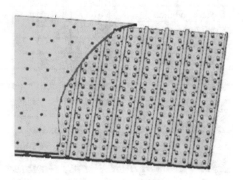

Luo et al. (2015) numerically investigated the effect of dimples on the heat transfer and friction factor in a Lamilloy cooling structure of a single-phase flow. Air jet impingement nozzle was positioned directly above the dimple. The presence of dimples on a Lamilloy cooling structure increased the heat transfer due to flow reattachment and recirculation. The dimple normalised depth and diameter were varied from 0 to 0.3 and 1 to 2.5, respectively.

The Reynolds number ranges from 10,000 to 70,000. With the increase of both normalised depth and normalised diameter of dimple, heat transfer on the targeted surface was first augmented due to increase in the flow reattachment and recirculation and then it was reduced due to large vortex ring. Double-wall cooling structure was developed to reduce the thermal stresses of gas turbine blades and improve the lifetime of engine.

Figures 2.11 and 2.12 show the double-wall cooling structure of turbine vane and Lamilloy cooling structure consisted of two layers, pin fins and film-cooling holes. Taylor and Hodge (2017) developed a model using discrete element method to evaluate heat transfer and pressure drop characteristics of fully developed turbulent

Fig. 2.13 Wolverine
Turbo-Chill tube (integral
fin outside surface and ten
start helical rib internal rib
roughness) (Webb and Kim
2005)

flow through pipes having three-dimensional roughnesses. They compared their results of numerical model with the experimental data of Dipprey and Sabersky (1963), Gowen and Smith (1968), Webb and Chamra (1991) and Rabas et al. (1993).

2.3.2 Integral Fin Tubes, Ribbed Tubes and Microchannels

In finned tubes, typically heat transfer occurs between gas and liquid, gas and two-phase liquid or sometimes between gases. The outside and tube-side enhancement surfaces increase the heat transfer characteristics. Figure 2.13 shows Wolverine Turbo-Chill. In this figure, the combination of integral fin outside and internally helical rib with rib roughness is presented. This generates higher heat transfer and hence is known as "doubly enhanced tube".

Typically lesser number of fins per unit length is used for boiling and condensation (19–26 fins/in.) in comparison to refrigeration (35 fins/in.). Moon and Kim (2013) used pin-fin arrays to enhance the heat transfer in the cooling channels of turbine blade. They reported that the uniformity in heat transfer and exit flow has increased with increase in pin-fin count. They have also presented the effect of space between the pin fins on heat transfer rates.

Figures 2.14 and 2.15 represent the effect of pin-fin count and pitch on performance functions. The performance functions are F_{Nu} (performance function based on Nusselt number), F_f (performance function based on frictional loss), F_η (thermal performance), and F_m (performance function based on mass flow rate). The pressure drop characteristics in terms of friction factor for smooth tubes and tubes having micro-fins have been presented by Cebi et al. (2013). They carried out a numerical investigation, compared their results with various available experimental data and reported minimal deviation. They have also presented various correlations available for friction factor, and also they have made calculations for smooth tubes and micro-fin tubes as shown in Tables 2.2 and 2.3, respectively. Song et al. (2017), Naresh et al. (2018) and Karami et al. (2018) also studied finned surfaces.

Chang and Wang (2017) presented correlations for heat transfer using multilouvered fins incorporated in brazed aluminium heat exchangers. Figure 2.16 shows the brazed aluminium heat exchanger. They carried out several experiments with 27 types of multilouvered fins. The geometrical parameters such as length and pitch of the louver, height and pitch of the fin and the tube length have been varied.

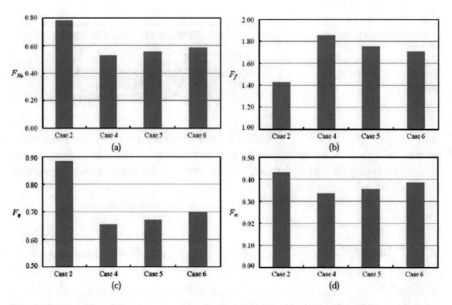

Fig. 2.14 The effect of pin-fin count on performance factors (Moon and Kim 2013)

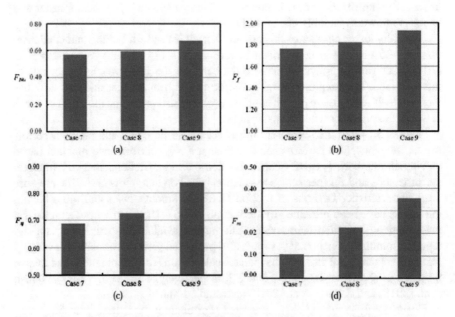

Fig. 2.15 The effect of pin-fin pitch on performance factors (Moon and Kim 2013)

Table 2.2 Correlations for friction factor in a smooth tube (Cebi et al. 2013)

Blasius (1913)	$f = 0.316 Re^{-1/4} (Re < 20{,}000)$ $f = 0.184 Re^{-1/5} (Re > 20{,}000)$
Petukhov (1970)	$f = (0.79 \ln Re - 1.64)^{-2}$
Danish et al. (2011)	$\dfrac{1}{2\sqrt{f}} = A - \dfrac{1.73718 A \ln A}{1.7378 + A} + \dfrac{2.62122 A (\ln A)^2}{(1.73718 + A)^3} + \dfrac{3.03568 A (\ln A)^3}{(1.7378 + A)^4}$ $A = 4\log Re - 0.4$
Churchill (1973)	$\dfrac{1}{\sqrt{f}} = -2\log \left\{ \left[\left(\dfrac{\varepsilon}{D}\right)^{3/7} \right] + (7/Re)^{0.9} \right\}$
Churchill (1977)	$f = 8 \left[\left(\dfrac{8}{Re}\right)^{12} + \dfrac{1}{(A+B)^{3/2}} \right]^{1/12}$ $A = \left(-2\log \left\{ \left[\left(\dfrac{\varepsilon/D}{3.7}\right) \right] + \left(\dfrac{7}{Re}\right)^{0.9} \right\} \right)^{16}$ $B = \left(\dfrac{37530}{Re} \right)^{16}$
Hrycak and Andruskhiw (1974)	$f = 3.1 \times 10^{-3} + 7.125 \times 10^5 Re - 9.7 \times 10^{10}$
Manadilli (1997)	$\dfrac{1}{\sqrt{f}} = -2\log \left(\dfrac{\varepsilon/D}{3.7} + \dfrac{95}{Re^{0.983}} - \dfrac{96.82}{Re} \right)$
Swamee and Jain (1976)	$f = \dfrac{0.25}{\left(\log\{ [(\varepsilon/D >)/3.7] + (5.74/Re^{0.9}) \} \right)^2}$
Moody (1944)	$f = 0.184 \, Re^{-1/5}$
Sonnad and Goudar (2006)	$\dfrac{1}{\sqrt{f}} = 0.8686 \ln \left[\dfrac{0.4587 Re}{S^{S/(S+1)}} \right]$
Easby (1978)	$\dfrac{f}{f_a} = 1.006 - 5.13 \dfrac{Gr}{Re^2}$
Bhatti and Shah (1987)	$f = A + \dfrac{B}{Re^{1/m}}$ $Re < 2100 \rightarrow A = 0, B = 16, m = 1$ $2100 < Re < 4000 \rightarrow A = 0.0054, B = 2.3 \times 10^{-3}, m = -\dfrac{2}{3}$ $Re > 4000 \rightarrow A = 1.28 \times 10^{-3}, B = 0.1143, m = 3.2154$

The geometrical parameters of the multilouvered fin are shown in Fig. 2.17. They reported an error of 10% after testing 85% of their experimental data with the correlations.

Suga and Aoki (2017) carried out a numerical study to design optimum geometry of multilouvered fins with optimum geometry. They recognised that the control of thermal wakes after louvers was the crucial issue for design optimisation. A correlation for various geometrical parameters has also been presented. The louver angle in the range of 20° and 30° showed the optimum performance of heat transfer enhancement.

Table 2.3 Correlations for friction factor in micro-fin tubes (Cebi et al. 2013)

Siddique and Alhazmy (2008)	$f = 0.907Re^{0.286}$
Al-Fahed et al. (1993)	$f = 0.9978Re^{0.2943}$
Zdaniuk et al. (2008)	$f = 0.128Re^{-0.305}n^{0.235}\left(\dfrac{e}{D}\right)^{0.319}\alpha^{0.397}$
Carvanos (1979)	$f = 0.046Re^{-0.2}\left(\dfrac{d}{d_{\mathrm h}}\right)^{1/2}\left(\dfrac{A}{A_{\mathrm n}}\right)^{0.5}(\sec\alpha)^{0.75}$
Ravigururajan and Bergles (1996)	$\dfrac{f}{f_{\mathrm{FI}}}\left(1+\left\{29.1Re^{a_1}\left(\dfrac{h}{d}\right)^{a_2}\left[\left(\dfrac{p}{d_{\mathrm h}}\right)\right]^{a_3}\left(\dfrac{\alpha}{90}\right)^{a_4}(1+1.47\cos\theta)\right\}^{16/15}\right)^{16/15}$ $$f_{\mathrm{FI}} = (1.58\ln Re - 3.28)^{-2}$$ $$a_1 = 0.67 - \frac{0.06p}{d} - \frac{0.49\alpha}{90}$$ $$a_2 = 1.37 - \frac{0.157p}{d}$$ $$a_3 = -1.66\times 10^{-6}Re - \frac{0.33\alpha}{90}$$ $$a_4 = 4.59 + 4.11\times 10^{-6}Re - \frac{0.15p}{d}$$
Wang et al. (1996)	$$\sqrt{\frac{f}{2}} = \frac{e^+}{Re}\left(\frac{D_{\mathrm i}}{e}\right)$$ $$e^+ < 23 \rightarrow e^+ = 0.4107 + 0.093675X_{\mathrm f} + \frac{0.58464}{\ln X_{\mathrm f}}$$ $$e^+ > 23 \rightarrow e^+ = 0.07313 + 0.09511X_{\mathrm f}$$ $$X_{\mathrm f} = Re\left(\frac{e}{D_{\mathrm i}+\lambda}\right)\frac{n^{0.25}}{(\cos\alpha)^{0.5}}$$
Haaland (1983)	$$f = \frac{0.3086}{\left\{\log\left[\frac{6.9}{Re} + \left(\frac{\varepsilon}{3.7D}\right)^{1.11}\right]\right\}^2}$$
Goto et al. (2001)	$f = 1.47\times 10^{-4}Re^{0.53}(2000 \leq Re \leq 2600)$ $f = 0.046Re^{0.2}(2600 < Re \leq 6500)$ $f = 1.23\times 10^{-2}Re^{0.21}\ (6500 \leq Re \leq 12{,}500)$ $f = 9.2\times 10^{-3}(12{,}500 < Re)$

Fig. 2.16 A brazed aluminium heat exchanger (Chang and Wang 2017)

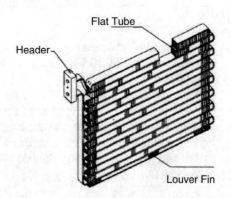

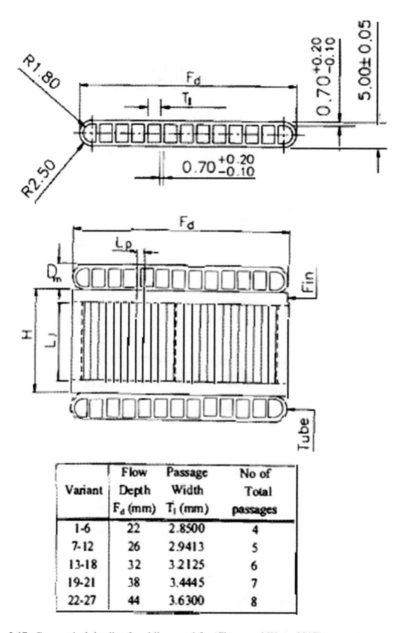

Fig. 2.17 Geometrical details of multilouvered fin (Chang and Wang 2017)

Variant	Flow Depth F_d (mm)	Passage Width T_l (mm)	No of Total passages
1-6	22	2.8500	4
7-12	26	2.9413	5
13-18	32	3.2125	6
19-21	38	3.4445	7
22-27	44	3.6300	8

Wang (2017) presented a report on the patents of various geometries and patterns of fins for fin and tube type of heat exchangers. They have compared nearly 51 models which were given patents during the period 1981–1999. Figure 2.18 shows some enhanced surfaces which gives higher performance than normal plain fins.

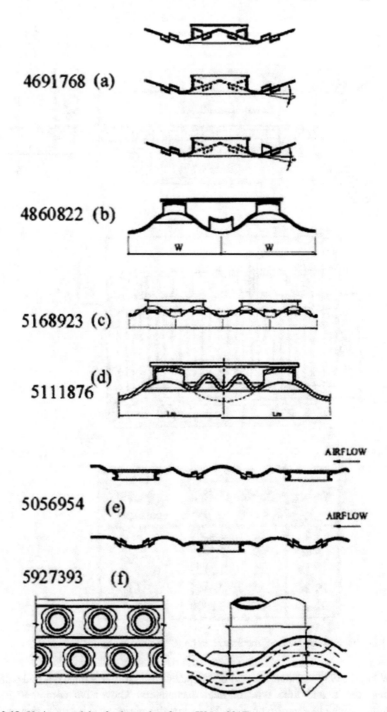

Fig. 2.18 Various models of enhanced surfaces (Wang 2017)

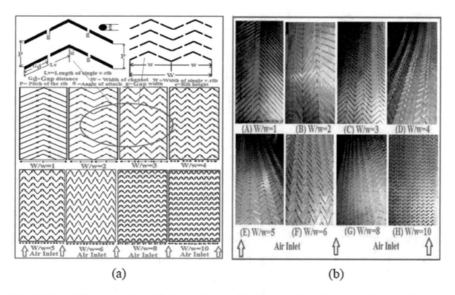

Fig. 2.19 (**a**) Discuss multiple V-ribs with gap, (**b**) photographic views of multiple V-rib gap (Kumar et al. 2014)

Table 2.4 Rib parameters corresponding to maximum, Nu/Nu_s (Kumar et al. 2014)

Rib parameters	Fixed parameters	Value of rib parameter	Maximum value of Nu/Nu_s
e/D	$W/w = 6$, $P/e = 10$, $Gd/Lv = 0.69$, $g/e = 1.0$, $\alpha = 60°$	0.043	6.45
W/w	$P/e = 8$, $e/D = 0.043$, $Gd/Lv = 0.69$, $g/e = 1.0$, $\alpha = 60°$	6.0	6.69
Gd/lv	$W/w = 6$, $P/e = 10$, $e/D = 0.043$, $g/e = 1.0$, $\alpha = 60°$	0.69	6.45
g/e	$W/w = 6$, $P/e = 10$, $e/D = 0.043$, $Gd/Lv = 0.69$, $\alpha = 60°$	1.0	6.45
α	$W/w = 6$, $P/e = 10$, $e/D = 0.043$, $Gd/Lv = 0.69$, $g/e = 1.0$	60°	6.45
P/e	$W/w = 6$, $e/D = 0.043$, $Gd/Lv = 0.69$, $g/e = 1.0$, $\alpha = 60°$	8	6.69

The enhancement of heat transfer in a continuous multi-V-rib with a gap of solar air channel was studied by an experimental investigation performed by Kumar et al. (2014). Figure 2.19a and b shows the multiple V-ribs with gap. Thermohydraulic performance was carried out in a single-phase convective mode of heat transfer by passive enhancement technique. Experiment was conducted on roughened surface having different roughness geometry to collect heat transfer and friction data.

Table 2.4 presents the values of rib parameters for which the maximum value of Nu/Nu_s has been obtained. They observed that values of thermohydraulic performance increase with increase in Reynolds number and become almost asymptotic in the higher range of Re. Table 2.5 shows values of roughness geometry parameters for which f/f_s values have been found to be maximum. Han et al. (1989) and

Table 2.5 Rib parameters corresponding to maximum f/f_s (Kumar et al. 2014)

Rib parameters	Fixed parameters	Value of rib parameter	Maximum value of f/f_s
e/D	$W/w = 6$, $P/e = 10$, $Gd/Lv = 0.69$, $g/e = 1.0$, $\alpha = 60°$	0.043	5.49
Gd/Lv	$W/w = 6$, $P/e = 10$, $e/D = 0.043$, $g/e = 1.0$, $\alpha = 60°$	0.69	5.49
g/e	$W/w = 6$, $P/e = 10$, $e/D = 0.043$, $Gd/Lv = 0.69$, $\alpha = 60°$	1.0	5.49
α	$W/w = 6$, $P/e = 10$, $e/D = 0.043$, $Gd/Lv = 0.69$, $g/e = 1.0$	60°	5.49
P/e	$W/w = 6$, $e/D = 0.043$, $Gd/Lv = 0.69$, $g/e = 1.0$, $\alpha = 60°$	8.0	6.72
W/w	$P/e = 8$, $e/D = 0.043$, $Gd/Lv = 0.69$, $g/e = 1.0$, $\alpha = 60°$	10	7.03

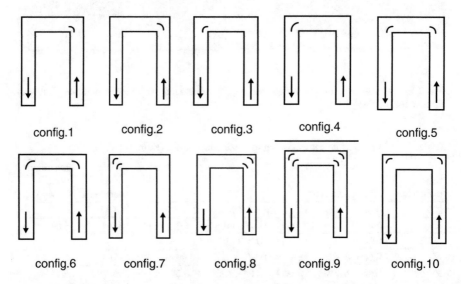

Fig. 2.20 Tested configuration of U-bend channel (from Cimina et al. 2015)

Taslim et al. (1996) experimentally studied the fully developed flow on heat transfer and friction factor in an air channel roughened with various ribs.

Saini and Saini (1997) and Kiml et al. (2001) investigated an indoor and outdoor experimental work by using roughened wall and roughened rib, respectively, and studied the effect of roughened surface on Stanton number and pressure. Cimina et al. (2015) introduced a heat transfer technique in a single-phase flow by using various guide vanes and ribs in a U-bend channel. Transverse and 45° V-shaped ribs were attached with guide vanes at outer wall to improve further the rate of heat transfer.

Figure 2.20 shows the details of the configurations of U-bend channel. Figure 2.21 illustrates the thermal performance for different ribbed wall with guide vanes. Among all the combinations of guide vanes and ribs, 45° V-shaped downstream

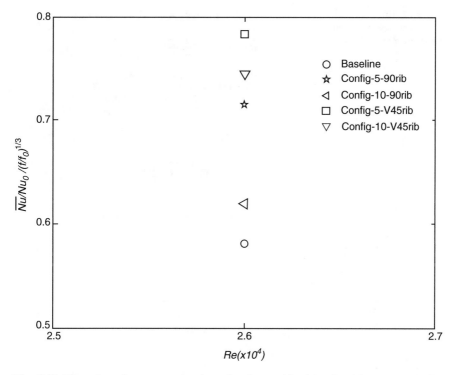

Fig. 2.21 Thermal performance comparison for the combination of guide vanes and ribs (Cimina et al. 2015)

pointing ribs coupled with configuration 5 gave the highest thermal performance. Their data can be utilised to improve the internal design of modern gas turbine blade cooling system.

Kumar et al. (2019) experimentally studied flow through square duct with a nanofluid and protrusion ribs incorporated in it. They discussed the nanofluid preparation and various thermophysical properties of nanofluids. They observed that the Nusselt number and friction factor for the rough duct with ribs and nanofluid increased with increase in volume fraction of the nanoparticles and decreased with decrease in diameter of the nanoparticles. The values of Nusselt number and friction factor have been noted to be the maximum at a volume fraction of 4% and with particle diameter of 30 μm. Figure 2.22 shows the effect of pitch and rib height of the protrusion. The overall enhancement of 2.37 has been observed using the compound technique for duct flow.

Krishna et al. (2018) conducted numerical analysis on heat transfer characteristics of flow through wavy microchannels. The relative waviness and aspect ratio have been varied to investigate their effect on heat transfer enhancement. Figures 2.23 and 2.24 show the variation of Nusselt number and friction factor with Reynolds number, respectively, for both wavy microchannel and straight

Fig. 2.22 Effect of pitch and height of a protruding rib (Kumar et al. 2019)

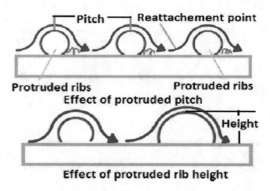

Effect of protruded pitch

Effect of protruded rib height

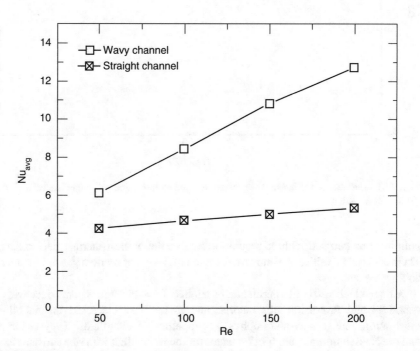

Fig. 2.23 Variation of Nusselt number with Reynolds number for wavy and straight microchannels (Krishna et al. 2018)

microchannel. It can be clearly seen that, although both Nusselt number and friction factor have increased for wavy microchannels, the increase in Nusselt number is quite significant as compared to that in friction factor. The increase in Nusselt number was noted to be 135% and the increase in friction factor was 35% over those obtained for straight microchannels in laminar flow regime. They explained that the enhancement occurred due to the vortices developed in the microchannels. They have also reported the performance of hybrid microchannels to be lower than that of wavy microchannels.

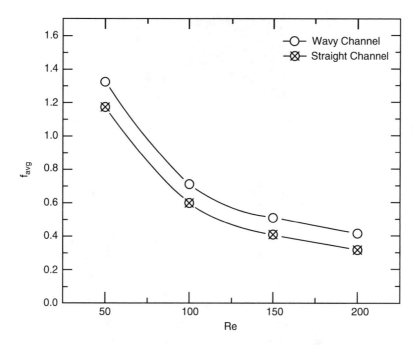

Fig. 2.24 Variation of average friction factor with Reynolds number for wavy and straight microchannels (Krishna et al. 2018)

The performance of microchannels with higher waviness was observed to be better. The use of nanofluids in microchannels acting as sink, to enhance the cooling, has been experimentally studied by Sandhu et al. (2018). Different ratios of water and ethylene glycol have been used as base fluids, while alumina nanoparticles have been dispersed in them to form nanofluids. For different nanofluids used, the enhancement has been observed to be in the range of 6.2–36%. They have also reported very less power required to pump the nanofluid.

2.3.3 Enhanced Condensing Tubes

These tubes have sawtooth fin geometry and are specially designed for increasing condensation coefficients in comparison to the standard integral fin tubes. Figure 2.25 shows the tube named as Sumitomo Tred-19D. Beaini and Carey (2013) presented a state-of-the-art review on enhancing dropwise condensation heat transfer. They suggested certain aspects which are to be taken care of for heat transfer augmentation such as thermal resistance for textured surfaces, decrease in heat transport for droplets having quite small diameter and measurement of contact angle. Figure 2.26 shows the dropwise condensation on an untreated copper surface with mirror finish. Figure 2.27 shows the scanning electron microscopic view of mirror-finished copper surface.

Fig. 2.25 Tubes for refrigerant condensation having enhancement on the condensing refrigerant side (outside) and the water side (inside), Sumitomo Tred-19D (Webb and Kim 2005)

Fig. 2.26 Dropwise condensation on an untreated copper surface with mirror finish (Beaini and Carey 2013)

Fig. 2.27 Scanning electron microscope view of mirror-finished copper surface (Beaini and Carey 2013)

Fig. 2.28 Sessile drop contact angle of 86° (Beaini and Carey 2013)

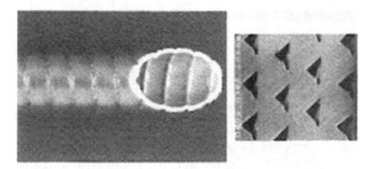

Fig. 2.29 Wolverine Turbo-B enhanced boiling tube (Webb and Kim 2005)

The sessile drop contact angle of 86° has been shown in Fig. 2.28. Kuzma-Kichta and Leontiev (2018) thoroughly reviewed the various macroscale, microscale and nanoscale heat and mass transfer literature and presented a detailed review which mainly sheds light on condensation heat transfer and boiling heat transfer. Kumar et al. (2002), Boyd et al. (1983), Baisar and Briggs (2009), Murase et al. (2006) and Zhang et al. (2012) worked on condensation heat transfer using various passive techniques.

2.3.4 Enhanced Boiling Tubes

Figure 2.29 shows Wolverine Turbo-B. The speciality of tube is that it works at much lower heat fluxes, and it boosts up nucleate boiling which does not occur at low heat flux in integral fin tubes. Terekhov et al. (2018) numerically investigated heat and mass transfer characteristics of evaporative cooling of airflow between two parallel plates in a direct evaporating cooler. An increase in vapour concentration has been observed along with an increase in temperature of the air at the exit, on addition of heat flux. They have also presented an analogy between momentum loss, pressure drop and friction factor and heat transfer process.

Other works on evaporative cooling have been carried out by Haji and Chow (1988), Yan and Lin (1988), Wei-Mon (1992), Volchkov et al. (2004),

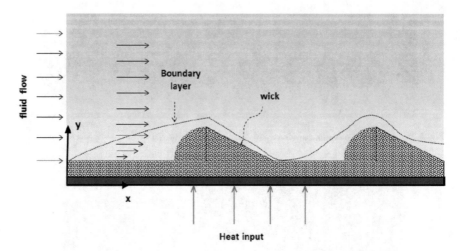

Fig. 2.30 Process of heat transfer in a microchannel having a porous wick (Carbajal et al. 2013)

Terekhov et al. (2015), Debbissi et al. (2008) and Nasr et al. (2009). Carbajal et al. (2013) numerically studied various porous surface structures having different geometries for boiling heat transfer enhancement. Four geometries, namely, flat, hemispherical, drop shaped and triangular porous surface structures, have been considered. They reported that the performance of hemispherical porous surface structure was superior to that of the other geometries considered.

Figure 2.30 represents process of heat transfer in a minichannel having a porous wick. Shatto and Peterson (2017) presented a review on the use of twisted-tape inserts for flow boiling heat transfer enhancement. They attributed the enhancement due to twisted tape to different factors such as secondary swirl flow, increase in flow velocity, decrease in the hydraulic diameter of the channel, fin effect of the tape and centrifugal acceleration of the flow. They concluded that the use of twisted tapes can increase the critical heat flux by retarding the dryout. Table 2.6 shows the details of various works carried on flow boiling enhancement using twisted tapes. The early research details of flow boiling in micro-fin tubes have been shown in Table 2.7. Various studies on pool boiling with surfactants and polymer additives have been listed in Table 2.8.

Wen et al. (2015) dealt with the heat transfer technique on two-phase flow and boiling heat transfer characteristics of R600A (Isobutane). They used perforated copper porous inserts in an annular tube. They examined the influence of heat flux, vapour quality, mass flux and insert geometry on heat transfer and pressure drop. The detail information of physical parameters and dimensions of porous channels is summarised in Table 2.9. Hsieh et al. (2000) conducted an experiment to enhance the heat transfer by using longitudinal strips with and without perforated holes and cross-strip inserts in a horizontal plain tube on isobutane two-phase flow boiling.

Table 2.6 Different studies on flow boiling enhancement using twisted tapes (Shatto and Peterson 2017)

Reference	Fluid	Tape twist	Tube ID (mm)
Viskanta (1961)	Water	5, 10	7.94
Gambill et al. (1960)	Water	2.3–12.0	3.45–6.33
Blatt and Adt (1963)	R-11, water	2.5, 5.0, 7.5	3.81, 6.35, 12.7
Gambill (1965)	Water	3.0, 4.9, 7.2, 12.0, 21.4	6.86
Matzner (1965)	Water	5, 15	10.2
Lopina and Bergles (1967)	Water	2.48–9.2	4.92
Bergles et al. (1971)	Nitrogen	4.1, 8.5	10.2
Sephton (1971)	Water	6.95	50.8 (OD)
Lopina and Bergles (1973)	Water	2.48–9.2	5.03
Cumo et al. (1974)	R-12	4.4	7.56
Agrawal et al. (1982)	R-12	3.76, 5.58, 7.37, 10.15	10.0
Bensler (1984)	R-113	3.94, 8.94, 13.92	8.10
Reid (1986)	R-113	11.6	10.92

Table 2.7 Research details of flow boiling in micro-fins (Kandlikar and Raykoff 2017)

	Researcher	Year	Tube OD (mm)	Tube material	Refrigerant	Test setup[a]
1	Ito and Kimura	1979	12.7	Cu/Al	R-22	ER
2	Shinohara and Tobe[**]	1985	9.52		R-22	
3	Khanpara et al.	1986	9.52	Cu	R-113	ER
4	Khanpara et al.	1987	9.52	Cu	R-113	ER
5	Khanpara et al.[*]	1987	9.52	Cu	R-22/R-113	ER
6	Schlager et al.	1990	11.7/12.7	Cu	R-22	HE
7	Reid et al.[*]	1991	9.53	Cu	R-113	ER
8	Eckels et al.	1992	7.9/9.5	Cu	R-22	HE
9	Ha and Bergles[*]	1993	9.5	Cu	R-12/oil[d]	ER
10	Chiang	1993	7.5/10	Cu	R-22/1%oil	HE
11	Thors and Bogart	1994	9.5/15.9	Cu	R-22	HE
12	Chamra and Webb	1995	–	Cu	R-22	HE
13	Koyama et al.[*]	1995	10	Cu	R-22/R-134a/R-123	HE
14	Singh et al.	1996	12.7	Cu	R-134a	ER

Vasiliev et al. (2012) investigated the effect of porous coating of a heated wall on mini-scale heat transfer in the minichannel. The cylindrical heated tube is placed inside a transparent coaxial glass tube. It provided evaporation and two-phase convective heat transfer enhancement. Zhou et al. (2017), Piasecka and Maciejewska (2015), Prajapati et al. (2016), Zhai et al. (2017), Al-Zaidi et al. (2018), (Zhang et al. 2017a, b, c), Sohel et al. (2014) and Karayiannis et al. (2018) worked on minichannel and microchannels in recent years.

Table 2.8 Various studies on pool boiling heat transfer enhancement using surfactants and polymer additives (Wasekar and Manglik 2018)

Author(s)	Heater geometry and heat flux level (kW/m^2)	Additives	Effect on heat transfer
		Surfactant solutions:	
Morgan et al. (1949)	Cylinder 0–500	Drene (triethanolamine alkyl sulphate), and sodium lauryl sulphonate	Enhances heat transfer with a maximum of around 100%
Jontz and Myers (1960)	Plate 20–25	Tergitol (sodium tetradecyl sulphate), and aerosol-22 (n-octadecyl tetrasodium 1,2 dicarboxyethyl sulphosuccinamate)	Heat transfer coefficient enhancement to the extent of 50%
Podsushnyy et al. (1980)	Cylinder 0–80	PVS-6 polyvinyl alcohol, NP-3 sulphonol, and SV 1017 wetting agent	Enhancement has an optimum value for concentration corresponding to c.m.c. In the higher heat flux range, the enhancement is higher for larger size heater
Filippov and Saltanov (1982)	Cylinder 0–100	Octadecylamine	Heat transfer coefficient improved by 100%
Yang and Maa (1983)	Plate 0–600	SLS (sodium lauryl sulphate), and SLBS (sodium lauryl benzene sulphonate)	Maximum enhancement of 100–200% in the nucleate boiling regime, and a 50–100% increase in critical heat flux. Also, the enhancement depends on particular surfactant in solution
Saltanov et al. (1986)	Cylinder 40–120	Octadecylamine	Maximum enhancement to the extent of 100% for an optimum level of additive concentration
Tzan and Yang (1990)	Cylinder 0–400	SLS (sodium lauryl sulphate)	Enhanced with a maximum increase of around 200% with an optimum value of surfactant concentration at high heat fluxes (>300 kW/m^2)
Liu et al. (1990)	Plate 0–400	BA-1, BA-2, BA-3, BA-4, DPE-1, DPE-3. Gelatine, oleic acid, trimethyl octadecyl ammonium chloride, trialkyl methyl ammonium chloride, and polyvinyl alcohol	Maximum enhance item in die range of 200–700% observed with BA-1, BA-2 and BA-3, while no effect with other additives

(continued)

Table 2.8 (continued)

Author(s)	Heater geometry and heat flux level (kW/m^2)	Additives	Effect on heat transfer
Chou and Yang (1991)	Plate 0–250	SLS (sodium lauryl sulphate)	Maximum enhancement of around 150%
Wu and Yang (1992)	Cylinder 23	SLS (sodium lauryl sulphate)	Decrease in incipient superheat and reduction in bubble size
Wang and Hartnett (1994)	Wire 0–600	SLS (sodium lauryl sulphate), and Tween-80 (polyoxyethylene sorbitan mono-oleate)	Heat transfer performance with SLS is similar to that of pure water; with Tween-80, the performance is slightly inferior

Table 2.9 Physical parameters and dimensions of the porous channels (from Wen et al. 2015)

Tube no.	Diameter of copper. D_p (mm)	Diameter of hollow (mm)	Surface area density (mm^2/mm^3)	Mean pore diameter, D_e (mm)	Porosity	Permeability (m^2)
1	Open tube					
2	2	0.9	6.25	0.293	0.181	1.31E−09
3	2	0	3.0	0.137	0.093	0.02E−09
4	4	1.1	2.16	0.386	0.224	1.67E−09
5	5	1.5	1.81	0.475	0.263	4.63E−09

2.3.5 Inserts and Swirl Devices

Yakovlev (2013) studied the performance of continuous twisted annular channels which are used for heat transfer enhancement in single-phase flow of water. The other works carried out on annular flow heat transfer with continuous twisted tape have been shown in Table 2.10. They presented a comparison of the performance of concave and convex annular channel surfaces. The experimental data has been presented for both laminar and turbulent flow regimes. Fiebig (2017) studied the compact heat exchangers having different wing-type vortex generators. The schematic of tube-fin and plate-fin compact heat exchangers has been shown in Fig. 2.31.

Han et al. (2017) investigated heat transfer and pressure drop characteristics in flow through square ducts with wedge-shaped and delta-shaped turbulence promoters. The square duct surface had two opposite ribbed walls. They observed that the enhancement in square duct with delta-shaped inserts was superior to that with wedge-shaped inserts in the investigated Reynolds number range of 15,000–80,000. The comparison of heat transfer performance of ribs with different

Table 2.10 Different works on annular flow heat transfer enhancement using continuous twisted length (Yakovlev 2013)

Ref.	h (mm)	d_1/d_2	$2h/D$	Heat transfer agent	Conditions of heat transfer
Tarasov and Shchukin (1977)	1.5–4.5	0.327–0.781	0.192–0.909	Air ($P \approx 0.1$ MPa, $Re_h = 10^3$–5×10^4)	Cooling on concave surface
Vilemas and Poshkas (1992)	2–6	0.665–0.889	0.024–0.357	Air ($P \approx 0.1$ MPa. $Re_h = 10^3$–3.4×10^5)	Two-sided heating
Boltenko et al. (2001, 2007)	1	0.778	0.118–0.233	Water ($P \approx 10$ MPa, $Re_h = 2 \times 10^3$– 5×10^4)	Two-sided heating
Ustimenko (1977)	30–50	0.74–0.9	0.139	Air ≈ 0.1 MPa. $Re_h = 3 \times 10^4$– 1.5×10^5)	Two-sided heating

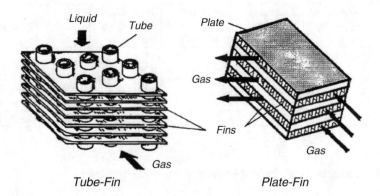

Fig. 2.31 Schematic of compact heat exchangers (Fiebig 2017)

geometrical parameters and designs has been presented in Table 2.11. Several researchers worked on tube inserts of several common types: coiled wire (San et al. 2015; Shafaee et al. 2016), spiral spring (Zhang et al. 2015), static mixer (Meng et al. 2016), vortex generator (Duangthongsuk and Wongwises 2013; Bali and Sarac 2014) and twisted tape (Bhuiya et al. 2013; Varun et al. 2016; Saysroy and Eiamsa-ard 2017).

Zhang et al. (2017c) studied the effect of using combined heat transfer enhancement technique with twisted tapes and interrupted ribs. The twisted tapes with three twist ratios are considered. The twisted tapes are placed in tubes with in-line ribs and staggered ribs. The tubes with twisted tapes and ribs are shown in Fig. 2.32. The Nusselt number and friction factor for the inserts and ribs are shown in Fig. 2.33. They reported that under constant pumping power, the tube having twisted tape and

Table 2.11 Performance comparison of ribs with different geometry and design (Han et al. 2017)

Turbulator configuration (surface area increment, %)	$Re = 15,000$		$Re = 30,000$		$Re = 50,000$		$Re = 80,000$	
	Nu_r/Nu_0	f/f_0	Nu_r/Nu_0	f/f_0	Nu_r/Nu_0	f/f_0	Nu_r/Nu_0	f/f_0
90° continuous rib (20%)	2.5	5.5	2.2	5.8	2.1	6.0	1.9	6.5
90° broken rib (23%)	3.2	8.0	2.9	8.1	2.5	8.4	2.3	8.8
60° continuous rib (23%)	3.1	7.4	2.7	7.6	2.5	8.6	2.3	9.6
60° broken rib (26%)	3.7	7.0	3.5	7.5	3.1	7.8	2.9	8.1
60° V-shaped continuous rib (23%)	3.5	8.2	2.9	8.8	2.7	9.7	2.5	10.8
60° V-shaped broken rib (28%)	4.2	6.5	3.7	7.0	3.2	7.3	3.0	7.5
Wedge-shaped continuous rib (24%)	2.3	8.8	2.2	9.1	2.0	10.0	1.9	10.7
Wedge-shaped broken rib (36%)	2.9	9.6	2.8	10.2	2.6	10.7	2.3	10.8
Delta-shaped forward aligned rib (22%)	3.0	6.3	2.9	7.5	2.8	8.5	2.5	8.8
Delta-shaped backward aligned rib (22%)	3.9	7.0	3.3	7.3	3.0	8.1	2.8	8.7

staggered ribs showed best heat transfer and pressure drop performance. A superior performance of 25–40% has been observed for the combination of twisted tape and staggered ribs over that of using twisted tape alone. The performance of sine wave fins as heat transfer promoters has been studied by Kim and Youn (2013). The sine wave fins are showed in Fig. 2.34.

Hamdan (2016) numerically studied the swirl generator in laminar flow to enhance the heat transfer. The results showed that decaying swirl velocity augmented local heat transfer, whereas it increased the pumping pressure. He observed that addition of a swirl generator in a tube heat exchanger minimised the size of heat exchanger. On the other hand, it increased the overall cost of the system. The data showed that the maximum local Nusselt number and the maximum local friction coefficient increased from no improvement to 40% and from no improvement to 360%, respectively, with increasing of swirl number from 0 to 2.5.

Swirl flow is a passive technique (Bergles 1998) since it does not require external power input. The experimental work (Manglik and Bergles 2002; Klepper 1972, and Algifri and Bhardwaj 1985) showed that short twisted tape improved the heat transfer. The experimental tests of Chang and Dhir (1995), Sara and Bali (2007), Eiamsa-Ard et al. (2009), Ayinde (2010), Yilmaz et al. (2003) and Steenbergen and Voskamp (1998) showed that swirl flow promotes heat transfer.

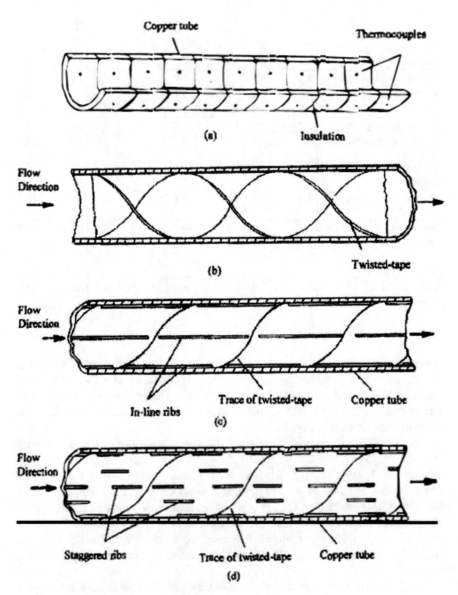

Fig. 2.32 Tubes with twisted tape and ribs (Zhang et al. 2017c)

2.3.6 Plate-Type Heat Exchangers

Figure 2.35 shows plate-type heat exchanger. Typically, corrugated surfaces are used to generate secondary flow and mixing. This increases heat transfer rate. Thonon et al. (2017) dealt with plate heat exchanger designing. The single-phase heat transfer, evaporation and condensation phenomena are considered.

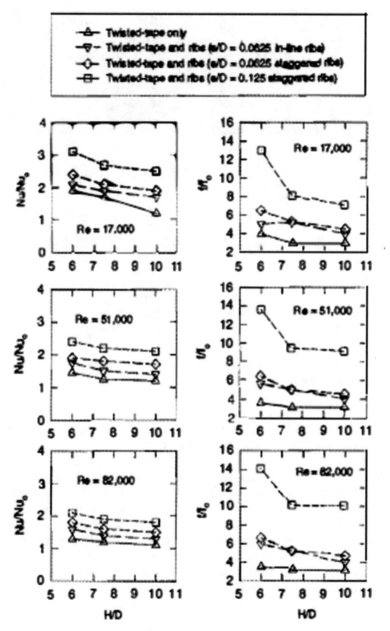

Fig. 2.33 Variation of Nusselt number and friction factor with twisted tape pitch-to-tube diameter ratio (H/D) (Zhang et al. 2017c)

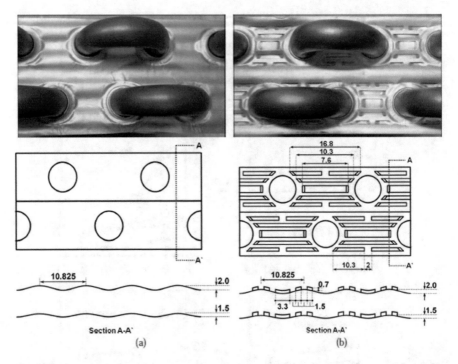

Fig. 2.34 Sine wave pictorial representation and drawings of (**a**) sine wave fins and (**b**) sine slit wave fins (Kim and Youn 2013)

Naik and Tiwari (2018) investigated the heat transfer and pressure drop characteristics of flow over a plate with circular cylinders placed on it. They mainly focused on the effect of aspect ratio of the cylinders on the thermohydraulic performance. They have mounted the cylinders on the plate in both in-line and staggered configurations. The variation of Nusselt number and friction factor with Reynolds number for in-line arrangement of cylinders having different aspect ratios is shown in Fig. 2.36a. Similarly, the performance of staggered arrangement is shown in Fig. 2.36b. The maximum enhancement of 73.16% over the plain plate is observed for in-line arrangement of cylinders, while the maximum enhancement was 87.79% for staggered arrangement. Also, the increase in enhancement with increase in aspect ratio is reported. The highest performance is observed for aspect ratio of 2. They concluded that the cylinders arranged in staggered position give much better enhancement.

Yang et al. (2016) presented a new model of plate-fin heat exchanger having a header with perforated wing panel. They carried out a numerical investigation to study the thermohydraulic performance of the plate-fin heat exchanger.

Cernecky et al. (2015) have studied experimentally convective heat transfer in horizontal plate heat exchange system. They focused on designing a heat exchanger surface, used in plate-type heat exchanger, heating or cooling equipment,

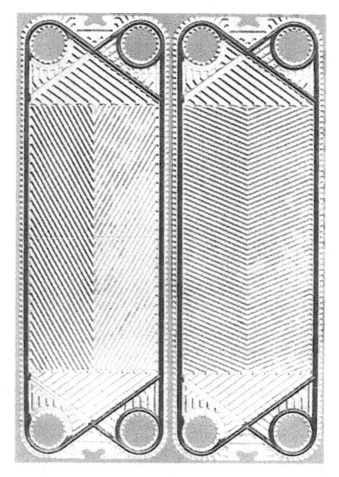

Fig. 2.35 Corrugated plate geometry used in plate-type heat exchangers (Webb and Kim 2005)

components in electronics, etc., for heat transfer at a low Reynolds number. Another objective of the experiment was to increase the heat transfer coefficient by installing directional tubing into the stream of forced airflow. The interferogram images of the heat exchange surfaces are shown in Fig. 2.37. A more detailed description of the interferogram analysis is given by Cernecky et al. (2014).

Table 2.12 gives the resulting values of local heat transfer coefficient (α_x) along the lower heated surfaces (X_1–X_5). Similarly, Table 2.13 shows the experimented values of local heat transfer coefficient along upper heated surfaces ($X_{1'}$–$X_{5'}$). The mean heat transfer coefficients on the lower surfaces α_m and upper surfaces $\alpha_{m'}$ are also presented in Tables 2.14 and 2.15. The experimental work revealed that an increase in the heat transfer is accompanied by increase in pressure loss.

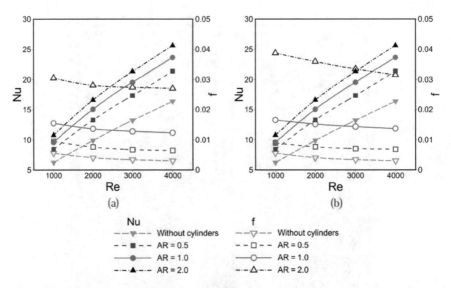

Fig. 2.36 Nusselt number and friction factor for (**a**) in-line arrangement and (**b**) staggered arrangement of cylinders having different aspect ratios (Naik and Tiwari 2018)

2.3.7   Other Relevant Works

Figure 2.38 shows special packing geometries for heat transfer enhancement. Fundamentally, evaporation due to small fraction of water causes cooling of water. This water circulated in tubes where heat exchanging process occurs. The special packing geometries contribute high heat and mass transfer coefficient to the flowing air. In cooling tower, heat and mass transfer occurs simultaneously. Figure 2.39 shows packaging for distillation columns. Typically it deals with heat and mass transfer between fluids, same as that of cooling tower.

Kuzenov and Ryzhkov (2018) developed a new mathematical model to study the convective heat and mass transfer enhancement in both laminar and turbulent regimes in the proximity of hypersonic aircraft. The proposed model could handle complex geometry of the aircraft even at low temperatures. The performance of reciprocating channels having rib installations for heat transfer augmentation is investigated by Perng and Wu (2013). The turbulent mixed convection is considered. They observed that the enhancement can be attributed to the effect of buoyancy forces, inertia forces and momentum of the inlet fluid. They concluded that the Nusselt number for the tube having circular ribs was in the range of 94.44–184.65% above that of the plain tube.

Zimparov et al. (2016) investigated the performance evaluation criteria of heat transfer surfaces in heat exchanger design. Figure 2.40 presents the variation of driving temperature difference with Reynolds number (upper part) and external thermal resistance (lower part) of the smooth and augmented tube. Table 2.16

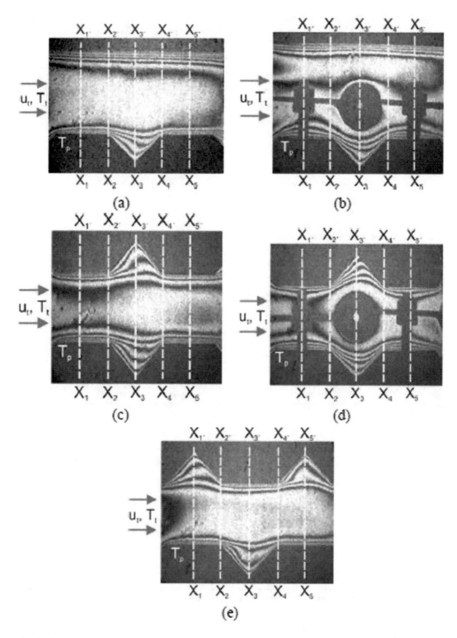

Fig. 2.37 Holographic interferogram images for heat exchanger system: (**a**) A1, (**b**) A2, (**c**) B1, (**d**) B2 and (**e**) C (Cernecky et al. 2015)

Table 2.12 Local and mean heat transfer parameters in investigated areas of heat exchange system A1, A2, B1, B2 and C-lower heated surfaces (from Cernecky et al. 2015)

Section	α_x [W/(m^2 K)]				
	A_1	A_2	B_1	B_2	C
X_1	13.22	4.83	9.98	8.75	16.44
X_2	12.75	6.47	8.95	12.23	18.41
X_3	2.61	2.77	2.68	4.27	3.03
X_4	13.99	6.70	14.05	12.34	18.15
X_5	12.26	5.33	14.82	10.48	20.62
α_m [W/(m^2 K)]	10.97	5.22	10.10	9.61	15.33
Nu_m [–]	25.41	12.09	20.28	19.29	30.78

Table 2.13 Local and mean heat transfer parameters in investigated areas of heat exchange system A1, A2, B1, B2 and C-lower heated surfaces (from Cernecky et al. 2015)

Section	α_x [W/(m^2 K)]				
	A_1	A_2	B_1	B_2	C
$X_{1'}$	6.09	9.92	10.37	13.00	3.18
$X_{2'}$	5.83	11.79	11.28	16.69	17.47
$X_{3'}$	5.53	12.48	2.69	4.14	15.91
$X_{4'}$	5.98	12.51	17.02	18.79	18.08
$X_{5'}$	5.75	10.63	16.46	14.88	3.02
$\alpha_{m'}$ [W/(m^2 K)]	5.84	11.47	11.56	13.50	11.53
$Nu_{m'}$ [–]	13.53	26.57	23.21	27.10	23.15

Table 2.14 Geometrical parameters of the smooth tube and wire-coil insert (from Zimparov et al. 2016)

No.	D_i, mm	p, mm	e, mm	p/e	e/D_i	p/D_i
WC	14.30	12.5	1.0	12.5	0.070	0.874

Table 2.15 Summary of cases studied

Case	β	α (first pass)	α (second pass)	D_h (mm)	Ω (rpm)/R_0	b_0	$\Delta\rho/\rho$	Expt.
Case 1	0°	–45°	+45°	13.2	550/+0.24	0.3	0.13	YES
Case 2	45°	–45°	+45°	13.2	550/+0.24	0.3	0.13	YES
Case 3	45°	–45°	+45°	13.2	–550/–0.24	0.3	0.13	NA
Case 4	45°	+45°	+45°	13.2	550/+0.24	0.3	0.13	NA

For all cases $Re = 25,000$, $P_{ref} = 10$ atm (from Brahim and Miloud 2016)

Fig. 2.38 Enhanced film-type cooling tower packing geometries (Webb and Kim 2005)

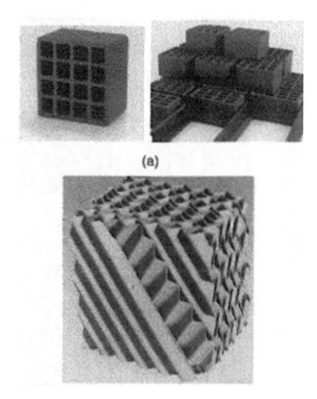

Fig. 2.39 Different packing for distillation columns (Webb and Kim 2005)

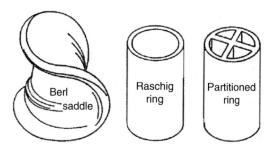

illustrates the geometrical parameters of the smooth tube and wire-coil insert. Figure 2.41 shows variation of experimental friction factor with different values of Reynolds number for smooth pipe.

Yilmaz et al. (2005) reviewed the performance evaluation criteria (PEC) for the design of heat exchanger based on analysis of first law of thermodynamics. Yilmaz et al. (2001) also investigated the PEC for heat exchanger design theory based on second law of thermodynamics. The entropy and exergy were used as evaluation parameters for calculating the PEC for heat exchanger design. Bejan (1978) gave a

Fig. 2.40 Variation of the
driving temperature and
external thermal resistance
with Reynolds number for
different Prandtl numbers
(Zimparov et al. 2016)

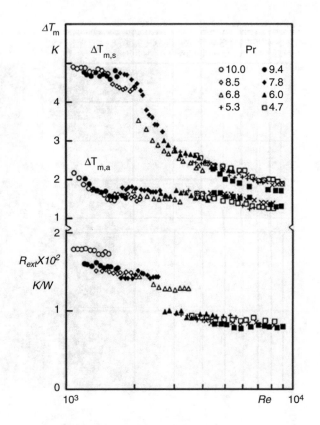

Table 2.16 Comparison of local mass flux for shower 1 at a variety of water pressure and fixed
shower-tip-to distance, $z = 70$ mm (Mishra et al. 2014)

Water pressure/tubes	1	2	3	4	5	Position
4.0 bar	56.21	79.15	179.23	68.12	48.28	Center
3.0 bar	51.40	72.95	165.81	66.32	48.08	
2.0 bar	34.82	33.16	149.23	39.79	36.47	

Fig. 2.41 Variation of the
Fanning friction factor with
Reynolds number
(Zimparov et al. 2016)

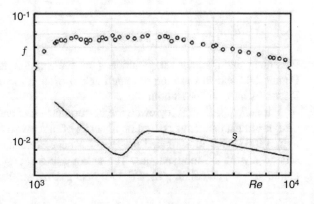

dimensionless number known as number of entropy production unit, N_s. Bejan (1982) gave the augmentation entropy generation number N_s, as a criterion to compare the rate of entropy generated in the heat exchanger before and after the use of heat transfer augmentation technique.

The special geometries used for optimisation of the design for heat exchanger are plate and frame with chevron plates (Muley and Manglik 2000), tube bundles with tape inserts (Yerra et al. 2007), rectangular twisted duct exchanger (Manglik et al. 2012), oval cross-section twisted tube exchanger (Bishara et al., 2013) and compact plate-fin and compact tube-fin heat exchanger (Manglik and Jog 2016). Brahim and Miloud (2016) investigated the effect of rotating channel and rib orientation on heat transfer enhancement of two-pass internal cooling of turbine blades.

Figure 2.42 shows the geometry of rib orientation, direction of rotation and cross section of blades from the axis of rotation. Two major forces, Coriolis and centrifugal forces, were observed which significantly affect the flow and heat transfer characteristics. Figure 2.43 illustrates these two major forces. Dutta et al. (1995) studied the effect of rib geometry in rotating two-pass square channel on heat transfer enhancement. Table 2.17 shows the detailed data of geometric parameters of ribs and channel in different cases as well as flow properties which are used in numerical simulation.

Griffith et al. (2002), Parsons et al. (2002), Azad et al. (2002), Iacovides et al. (1998) and Johnson et al. (1994) studied the effect of rib configuration on heat transfer in rotating channels. Mishra et al. (2014) investigated the heat transfer characteristics of shower cooling on a hot flat steel plate by using transient temperature measurement technique. It was observed that shower hole arrangement and design of shower head influenced the cooling rate at a higher temperature. Tables 2.18 and 2.19 show comparative data of local mass flux in five tubes for shower 1 and shower 2 at water pressure ranging from 2.0 to 4.0 bar and shower tip-to-plate distance of 70 mm. Shower 1 shows the general trend for local mass flux distribution, whereas shower 2 shows the opposite distribution. Results showed that shower design can be optimised to give optimal cooling.

The recent works on heat transfer enhancement are carried out by Wijayanta et al. (2017), Karatas and Derbentli (2017), Mahdavi et al. (2018), Hemadri et al. (2018), Hosseinalipour et al. (2018), Sharma et al. (2018), Nouri-Borujerdi and Nakhchi (2018), Leporini et al. (2018) and Mallor et al. (2019). These works in different areas of heat transfer enhancement are very important.

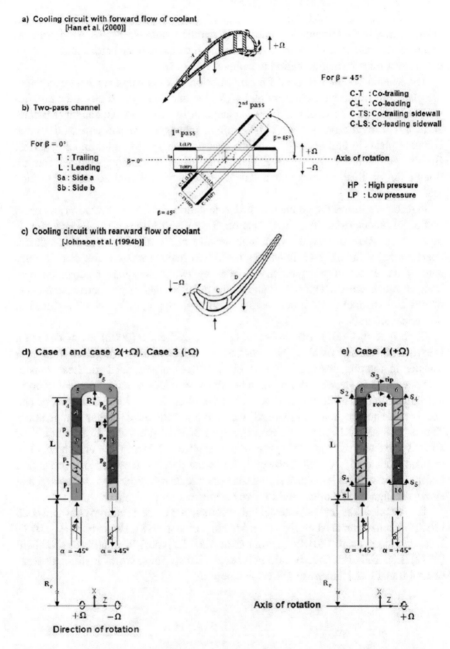

Fig. 2.42 Geometry of two-pass channel with rib orientation and rotation direction. Cross section of blades and channel viewed from axis of rotation (Brahim and Miloud 2016)

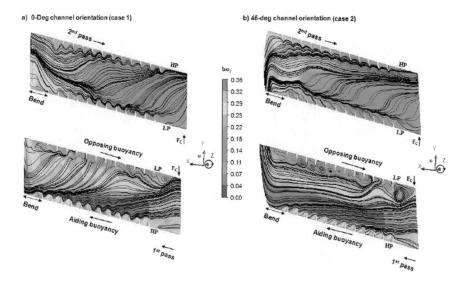

Fig. 2.43 Local buoyancy parameter contours and streamlines in symmetry plane in case 1 and in diagonal plane in case 2 (Brahim and Miloud 2016)

Table 2.17 Comparison of local mass flux for shower 1 at a variety of water pressure and fixed shower-tip-to distance, $z = 70$ mm (Mishra et al. 2014)

Water pressure/tubes	1	2	3	4	5	Position
4.0 bar	54.82	45.39	33.64	43.55	51.71	Center
3.0 bar	48.333	37.21	21.444	35.28	44.98	
2.0 bar	41.45	28.18	11.55	26.53	38.13	

Table 2.18 Various proposed correlations for the estimation of Nusselt number for nanofluid flowing in a tube with twisted and wire-coil inserts (Naik et al. 2014)

Nanofluid/(insert)	Expression	Range	Refs.
Al₂O₃-water (twisted tape)	$Nu = 0.03666Re^{0.8204}Pr^{0.4}(0.001+\varphi)^{0.04704}\left(0.001+\dfrac{H}{D}\right)^{0.06281}$	$10{,}000 < Re < 22{,}000$ $0{-}0.5\%$ $450 < Pr < 5.5$ $0 < H/D < 83$	Sundar and Sharma [27]
CuO-water (twisted tape)	$Nu = 0.005Re^{1.062}Pr^{0.5}(1+\varphi)^{0.112}$	$830 < Re < 1990$ $0.3{-}0.7\%, H/D = 3$	Wongcharee and Eiamsa-ard [29]
CuO-70:30% W/PG (twisted tape)	$Nu = 0.125Re^{0.5855}Pr^{0.4}(1+\varphi)^{0.3772}(1+H/D)^{0.05351}$	$1000 < Re < 10{,}000$ $0{-}0.5\%, 4.50 < Pr < 5.5$ $0 < H/D < 83$	Naik et al. [38]
Fe₃O₄-water (twisted tape)	$Nu = 0.0223Re^{0.8}Pr^{0.5}(1+\varphi)^{0.54}(1+H/D)^{0.028}$	$3000 < Re < 22{,}000$ $0{-}0.6\%$, $3.19 < Pr < 6.5$ $0 < H/D < 15$	Sudar et al. [50]
TiO₂-water	$Nu = 0.5He^{0.522}Pr^{0.613}\varphi^{0.0815}$	$500 < Re < 4500$	Kahani et al. [37]
Al₂O₃-water (wire coil)	$Nu = 0.7068He^{0.514}Pr^{0.563}\varphi^{0.112}$	0.25% and 0.1% $5.89 < Pr < 8.95$ $115.3 < He < 1311.4$	
CuO-base oil (wire coil)	$Nu = 0.467Re^{0.636}Pr^{0.324}\left(\dfrac{P}{d}\right)^{-0.358}\left(\dfrac{t}{d}\right)^{0.448}\left(\dfrac{\mu_y}{\mu_n}\right)^{-0.14}$	$20 < Re < 120$ $0{-}0.3\%$ $p/d = 1.79, 2.14$ and 2.50	Saeedinia et al. [36]
Al₂O₃-water (wire coil)	$Nu = 0.279(RePr)^{0.558}\left(\dfrac{P}{d}\right)^{-0.477}(1+\varphi)^{134.65}$	$Re < 2300 \; \varphi = 0.1\%$ $2 < p/d < 3$	Chadrasekar et al. [34]

Table 2.19 Various proposed correlations for the estimation of friction factor for nanofluid flowing in a tube with twisted and wire-coil inserts (Naik et al. 2014)

Nanofluid/(insert)	Expression	Range	Refs.
Al_2O_3-water (twisted tape)	$Nu = 0.03666Re^{0.8204}Pr^{0.4}(0.0001 + \varphi)^{0.04704}\left(0.0001 + \dfrac{H}{D}\right)^{0.06281}$	$10{,}000 < Re < 22{,}000$ 0–5%, $450 < Pr < 5.5$ $0 < H/D < 83$	Sundar and Sharma [27]
CuO-water (twisted tape)	$Nu = 0.005Re^{1.062}Pr^{0.5}(1 + \varphi)^{0.112}$	$830 < Re < 1990$ 0.3–0.7%, $H/D = 3$	Wongcharee and Eiamsaard [29]
CuO-70:30% W/PG (twisted tape)	$Nu = 0.125Re^{0.5855}Pr^{0.4}(1 + \varphi)^{0.3772}(1 + H/D)^{0.05551}$	$1000 < Re < 10{,}000$ 0–0.5%, $4.50 < Pr < 5.5$ $0 < H/D < 83$	Naik et al. [38]
Fe_3O_4-water (twisted tape)	$Nu = 0.0223Re^{0.8}Pr^{0.5}(1 + \varphi)^{0.54}(1 + H/D)^{0.028}$	$3000 < Re < 22{,}000$ 0–0.6%, $3.19 < Pr < 6.5$ $0 < H/D < 15$	Sudar et al. [50]
TiO_2-water	$Nu = 0.5He^{0.522}Pr^{0.613}\varphi^{0.0815}$	$500 < Re < 4500$	Kahani et al. [37]
Al_2O_3-water (wire coil)	$Nu = 0.7068He^{0.514}Pr^{0.563}\varphi^{0.112}$	0.25% and 0.1% $5.89 < Pr < 8.95$ $115.3 < He < 1311.4$	
CuO-base oil (wire coil)	$Nu = 0.4679\varphi^{0.636}Pr^{0.324}\left(\dfrac{P}{d}\right)^{-0.358}\left(\dfrac{P}{d}\right)^{0.448}\left(\dfrac{\mu_y}{\mu_n}\right)^{-0.14}$	$20 < Re < 120$ 0–0.3% $p/d = 1.79,\ 2.14$ and 2.50	Saeedinia et al. [36]
Al_2O_3-water (wire coil)	$Nu = 0.279(RePr)^{0.558}\left(\dfrac{P}{d}\right)^{-0.477}(1 + \varphi)^{134.65}$	$Re < 2300\ \varphi = 0.1\%$ $2 < p/d < 3$	Chadrasekar et al. [34]

References

Abraham S, Vedula RP (2016) Heat transfer and pressure drop measurements in a square cross-section converging channel with V and W rib turbulators. Exp Therm Fluid Sci 70:208–219

Agrawal KN, Varma HK, Lai S (1982) Pressure drop during forced convection boiling of R-12 under swirl flow. J Heat Transf 104:758–762

Al-Fahed SF, Ayub ZH, Al-Marafie AM, Soliman BM (1993) Heat transfer and pressure drop in a tube with internal microfins under turbulent water flow conditions. Exp Therm Fluid Sci 7 (3):249–253

Algifri A, Bhardwaj R (1985) Prediction of the heat transfer for decaying turbulent swirl flow in a tube. Int J Heat Mass Transfer 28(9):1637–1643

Al-Zaidi AH, Mahmoud MM, Karayiannis TG (2018) Condensation flow patterns and heat transfer in horizontal microchannels. Exp Therm Fluid Sci 90:153–173

Asako Y, Faghri M (1987) Finite-volume solutions for laminar flow and heat transfer in a corrugated duct. J Heat Transfer 109(3):627–634

Ayinde T (2010) A generalized relationship for swirl decay in laminar pipe flow. Sadhana 35 (2):129–137

Azad GMS, Uddin MJ, Han JC, Moon HK, Glezer B (2002) Heat transfer in two-pass rectangular rotating channels with 45 deg parallel and crossed rib turbulators. J Turbomach 124(2):251–259

Baisar M, Briggs A (2009) Condensation of steam on pin-fin tubes: effect of circumferential pin thickness and spacing. Heat Transfer Eng 30(13):1017–1023

Bali T, Sarac BA (2014) Experimental investigation of decaying swirl flow through a circular pipe for binary combination of vortex generators. Int Commun Heat Mass Transfer 53:174–179

Baxi CB, Ramachandran A (1969) Effect of vibration on heat transfer from spheres. J Heat Transfer 91(3):337–343

Beaini SS, Carey VP (2013) Strategies for developing surfaces to enhance dropwise condensation: exploring contact angles, droplet sizes, and patterning surfaces. J Enhanc Heat Transf 20 (1):33–42

Bejan A (1978) General criterion for rating heat-exchanger performance. Int J Heat Mass Transf 21:655–658

Bejan A (1982) Entropy generation through heat and fluid flow. Wiley, New York

Bensler HP (1984) Saturated forced convective boiling heat transfer with twisted tape inserts, Master of Science thesis, The University of Wisconsin-Milwaukee

Bergles AE, Brown Jr GS, Snider WD (1971) Heat transfer performance of internally finned tubes, ASME paper no. 71-HT-31

Bergles AE (1998) Techniques to enhance heat transfer. Handb Heat Transfer 3:11–11

Bhatti MS, Shah RK (1987) Turbulent and transition flow convective heat transfer in ducts. In: Kakac S, Shah RK, Aung W (eds) Handbook of single-phase convective heat transfer, chapter 4. Wiley, New York, p 16

Bhuiya MMK, Chowdhury MSU, Shahabuddin M, Saha M, Memon LA (2013) Thermal charac-teristics in a heat exchanger tube fitted with triple twisted tape inserts. Int Commun Heat Mass Transfer 48:124–132

Bishara F, Jog MA, Manglik RM (2013) Heat transfer enhancement due to swirl effects in oval tubes twisted about their longitudinal axis. J Enhanc Heat Transf 20(4)

Blasius H (1913) Das aehnlichkeitsgesetz bei reibungsvorgängen in flüssigkeiten. In: Mitteilungen über Forschungsarbeiten auf dem Gebiete des Ingenieurwesens. Springer, Berlin, Heidelberg, pp 1–41

Blatt TA, Adt RR (1963) The effects of twisted tape swirl generators on the heat transfer rate and pressure drop of boiling Freon 11 and water. ASME Paper No. ASME-63-WA-42

Boltenko EA, Il'in GK, Tarasevich SE, Yakovlev AB (2007) Heat transfer in annular channels with flow twisting. Russ Aeronaut 50(3):287–291

Boltenko EA, Tarasevich SE, Obuhova LA (2001) Heat transfer intensification in annular channels with a flow twisting. A convective heat transfer. Izv Ross Akad Nauk Energetika (3):99–104

Boyd LW, Hammon JC, Littrel JJ, Withers JG (1983) Efficiency improvement at Gallatin Unit 1 with corrugated condenser tubes. Am Soc Mech Eng 105(12): 88

Brahim B, Miloud A (2016) Prediction of flow and heat transfer inside a two-pass rotating channel with angled ribbed surfaces. J Enhanc Heat Transf 23(2):109–136

Brignoni LA, Garimella SV (2000) Effects of nozzle-inlet chamfering on pressure drop and heat transfer in confined air jet impingement. Int J Heat Mass Transfer 43:1133–1139

Carbajal G, Sobhan CB, Peterson GP (2013) Symmetrical porous surfaces for boiling enhancement in mini-channels: effects on liquid pressure drop. J Enhanc Heat Transf 20(1):73–81

Carnavos TC (1979) Heat transfer performance of internally finned tubes in turbulent flow. In: Proceedings of 18th national heat transfer conference. ASME, San Diego, CA/New York, pp 61–67, August 6–8, 1979

Cebi A, Celen A, Dalkilic AS, Wongwises S (2013) Friction factor characteristics for upward single-phase flows inside smooth and microfin tubes of a double-pipe heat exchanger for heating/cooling conditions. J Enhanc Heat Transf 20(5):413–425

Cernecky J, Koniar J, Brodnianska Z (2014) The effect of heat transfer area roughness on heat transfer enhancement by forced convection. J Heat Transf 136(4):041901

Cernecky J, Koniar J, Ohanka L, Brodnianska Z (2015) Temperature field and heat transfer in low Reynolds flows inside trapezoidal-profiled corrugated-plate channels. J Enhanc Heat Transf 22 (4):329–343

Chang F, Dhir V (1995) Mechanisms of heat transfer enhancement and slow decay of swirl in tubes using tangential injection. Int J Heat Fluid Flow 16(2):78–87

Chang SW, Yu KC (2018) Heat transfer enhancement by spirally coiled spring inserts with and without segmental solid cords. Exp Therm Fluid Sci 97:119–132

Chang YJ, Wang CC (2017) Brazed aluminum heat exchangers and their air side performance. J Enhanc Heat Transf 24(1–6):145–158

Chen W, Ren J, Jiang H (2011) Effect of turning vane configurations on heat transfer and pressure drop in a ribbed internal cooling system. ASME J Turbomach 133(4):041012

Chen XD, Xu XY, Nguang SK, Bergles AE (2001) Characterization of the effect of corrugation angles on hydrodynamic and heat transfer performance of fourstart spiral tubes. J Heat Transf 123:1149–1158

Chou CC, Yang YM (1991) Surfactant effects on the temperature profile within the superheated boundary layer and the mechanism of nucleate Pool boiling. J Chin Inst Chem Eng 22(2):71–80

Churchill SW (1973) Empirical expressions for the shear stress in turbulent flow in commercial pipe. AIChE J 19(2):375–376

Churchill SW (1977) Friction-factor equation spans all fluid-flow regimes. Chem Eng 84(24):91–92

Cimina S, Wang C, Wang L, Niro A, Sunden B (2015) Experimental study of pressure drop and heat transfer in a u-bend channel with various guide vanes and ribs. J Enhanc Heat Transf 22 (1):29–45

Cumo M, Farello GE, Ferrari G, Palazzi G (1974) The influence of twisted tapes in subcritical, once-through vapor generators in counter flow. J Heat Transfer 96(3):365–370

Danish M, Kumar S, Kumar S (2011) Approximate explicit analytical expressions of friction factor for flow of Bingham fluids in smooth pipes using Adomian decomposition method. Commun Nonlinear Sci Numer Simul 16(1):239–251

Debbissi C, Orfi J, Nassrallah S (2008) Numerical analysis of the evaporation of water by forced convection into humid air in partially wetted vertical plates. J Eng Appl Sci 3(11):811–821

Dipprey DF, Sabersky RH (1963) Heat and momentum transfer in smooth and rough tubes at various Prandtl numbers. Int J Heat Mass Transfer 6(5):329–353

Dizaji HS, Jafarmadar S (2014) Heat transfer enhancement due to air bubble injection into a horizontal double pipe heat exchanger. Int J Automot Eng 4(4):902–910

Duangthongsuk W, Wongwises S (2013) An experimental investigation of the heat transfer and pressure drop characteristics of a circular tube fitted with rotating turbine-type swirl generators. Exp Therm Fluid Sci 45:8–15

Dutta S, Han JC, Zhang YM (1995) Influence of rotation on heat transfer from a two-pass channel with periodically placed turbulence and secondary flow promoters. Int J Rotating Mach 1 (2):129–144

Easby JP (1978) The effect of buoyancy on flow and heat transfer for a gas passing down a vertical pipe at low turbulent Reynolds numbers. Int J Heat Mass Transfer 21(6):791–801

Eiamsa-Ard S, Thianpong C, Eiamsa-Ard P, Promvonge P (2009) Convective heat transfer in a circular tube with short-length twisted tape insert. Int Commun Heat Mass Transfer 36 (4):365–371

Elison B, Webb BW (1994) Local heat transfer to impinging liquid jets in the initially laminar, transitional, and turbulent regimes. Int J Heat Mass transfer 37(8):1207–1216

Fan CS, Metzger DE (1987, May) Effects of channel aspect ratio on heat transfer in rectangular passage sharp 180-deg turns. In: 32nd International gas turbine conference and exhibition

Fiebig M (2017) Compact heat exchangers: vortex generators. J Enhanc Heat Transf 24(1–6):1–20

Filippov GA, Saltanov GA (1982) Steam-liquid media heat-mass transfer and hydrodynamics with surface-active substance additives. Heat Transfer 4:443–447

Fukiba K, Ota K, Harashina Y (2018) Heat transfer enhancement of a heated cylinder with synthetic jet impingement from multiple orifices. Int Commun Heat Mass Transfer 99:1–6

Funfschilling D, Li HZ (2006) Effects of the injection period on the rise velocity and shape of a bubble in a non-Newtonian fluid. Chem Eng Res Des 84(10):875–883

Gambill WR (1965) Subcooled swirl-flow boiling and burnout with electrically heated twisted tapes and zero wall flux. J Heat Transfer 87(3):342–348

Gambill WR, Bundy RD, Wansbrough RW (1960) Heat transfer, burnout, and pressure drop for water in swirl flow through tubes with internal twisted tapes. Oak Ridge National Lab, TN

Garimella SV, Nenaydykh B (1996) Nozzle-geometry effects in liquid jet impingement heat transfer. Int J Heat Mass Transfer 39:2915–2923

Garimella SV, Rice RA (1995) Confined and submerged liquid jet impingement heat transfer. J Heat Transfer 117:871–877

Goldstein RJ, Timmers JF (1982) Visualization of heat transfer from arrays of impinging jets. Int J Heat Mass transfer 25(12):1857–1868

González-Altozano P, Gasque M, Ibáñez F, Gutiérrez-Colomer RP (2015) New methodology for the characterisation of thermal performance in a hot water storage tank during charging. Appl Therm Eng 84:196–205

Goto M, Inoue N, Ishiwatari N (2001) Condensation and evaporation heat transfer of R410A inside internally grooved horizontal tubes. Int J Refrig 24(7):628–638

Gowen RA, Smith JW (1968) Turbulent heat transfer from smooth and rough surfaces. Int J Heat Mass Transfer 11(11):57–1673

Griffith TS, Al-Hadhrami L, Han JC (2002) Heat transfer in rotating rectangular cooling channels (AR = 4) with angled ribs. J Heat Transfer 124(4):617–625

Gupta NK, Tiwari AK, Ghosh SK (2018) Heat transfer mechanisms in heat pipes using nanofluids—a review. Exp Therm Fluid Sci 90:84–100

Haaland SE (1983) Simple and explicit formulas for the friction factor in turbulent pipe flow. J Fluids Eng 105(1):89–90

Haji M, Chow L (1988) Experimental measurement of water evaporation rates into air and superheated steam. ASME J Heat Transf 110:237–242

Hamdan MO (2016) Numerical analysis of enhanced heat transfer in developing laminar pipe flow using decaying swirl at the inlet. J Enhanc Heat Transf 23(4):283–298

Han JC, Huang JJ, Lee CP (2017) Heat transfer in square channels with wedge-shaped and delta-shaped turbulence promoters. J Enhanc Heat Transf 24(1–6):101–116

Han JC, Park CK, Lei (1989) Augmented heat transfer in rectangular channels of narrow aspect ratios with rib turbulators. Int J Heat Mass Transfer 32:1619–1630

Hemadri V, Biradar GS, Shah N, Garg R, Bhandarkar UV, Agrawal A (2018) Experimental study of heat transfer in rarefied gas flow in a circular tube with constant wall temperature. Exp Therm Fluid Sci 93:326–333

Hosseinalipour SM, Shahbazian HR, Sunden B (2018) Experimental investigations and correlation development of convective heat transfer in a rotating smooth channel. Exp Therm Fluid Sci 94:316–328

Hrycak P, Andruskhiw R (1974) Calculation of critical Reynolds number in round pipes and infinite channels and heat transfer in transition regions. Heat Transfer 2:183–187

Hsieh SS, Jang KJ, Tsai YC (2000) Evaporation heat transfer and pressure drop in horizontal tubes with strip-type inserts using refrigerant 600a. J Heat Transfer 122(2):387–391

Iacovides H, Jackson DC, Ji H, Kelemenis C, Launder BE, Nikas K (1998) LDA study of the flow developing through an orthogonally rotating U-bend of strong curvature and rib roughened walls. ASME J Turbomach 120(2):386–391

Isaev SA, Leontiev AI, Chudnovsky Y, Popov I (2018) Vortex heat transfer enhancement in narrow channels with a single oval-trench dimple oriented at different angles to the flow. J Enhanc Heat Transf 25(6):565–577

Johnson BV, Wagner JH, Steuber GD, Yeh FC (1994) Heat transfer in rotating serpentine passages with trips skewed to the flow. ASME J Turbomach 116(1):113–123

Jontz PD, Myers JE (1960) The effect of dynamic surface tension on nucleate boiling coefficients. AIChE J 6(1):34–38

Kandlikar SG, Raykoff T (2017) Flow boiling heat transfer of refrigerants in microfin tubes. J Enhanc Heat Transf 24(1–6):231–242

Kang YT, Stout R, Christensen RN (1997) The effects of inclination angle on flooding in a helically fluted tube with a twisted insert. Int J Multiphase Flow 23(6):1111–1129

Karami M, Yaghoubi M, Keyhani A (2018) Experimental study of natural convection from an array of square fins. Exp Therm Fluid Sci 93:409–418

Karatas H, Derbentli T (2017) Three-dimensional natural convection and radiation in a rectangular cavity with one active vertical wall. Exp Therm Fluid Sci 88:277–287

Karayiannis T, Al-Zaidi AH, Mahmoud MM (2018) Condensation flow patterns and heat transfer in horizontal microchannels. Exp Therm Fluid Sci 90:153–173

Kareem ZS, Jaafar MM, Lazim TM, Abdullah S, Abdulwahid AF (2015) Passive heat transfer enhancement review in corrugation. Exp Therm Fluid Sci 68:22–38

Khoshvaght-Aliabadi M, Jafari A, Sartipzadeh O, Salami M (2016) Thermal–hydraulic performance of wavy plate-fin heat exchanger using passive techniques: perforations, winglets, and nanofluids. Int Commun Heat Mass Transf 78:231–240

Kim NH, Youn B (2013) Airside performance of fin-and-tube heat exchangers having sine wave or sine wave-slit fins. J Enhanc Heat Transf 20(1):43–58

Kiml R, Mochizuki S, Murata A (2001) Effects of rib arrangements on heat transfer and flow behavior in a rectangular rib-roughened passage: application to cooling of gas turbine blade trailing edge. J Heat Transfer 123(4):675–681

Kitagawa A, Kitada K, Hagiwara Y (2010) Experimental study on turbulent natural convection heat transfer in water with sub-millimeter-bubble injection. Exp Fluids 49(3):613–622

Klepper O (1972) Heat transfer performance of short twisted tapes. Oak Ridge National Laboratories, Oak Ridge, TN

Krishna PM, Deepu M, Shine SR (2018) Numerical investigation of wavy microchannels with rectangular cross section. J Enhanc Heat Transf 25(4–5):293–313

Kumar A, Chamoli S, Kumar M, Singh S (2016) Experimental investigation on thermal performance and fluid flow characteristics in circular cylindrical tube with circular perforated ring inserts. Exp Therm Fluid Sci 79:168–174

Kumar A, Chauhan R, Kumar R, Singh T, Sethi M, Sharm A (2017) Developing heat transfer and pressure loss in an air passage with multi discrete V-blockages. Exp Therm Fluid Sci 84:266–278

Kumar A, Saini RP, Saini JS (2014) An experimental investigation of enhanced heat transfer due to a gap in a continuous multiple V-rib arrangement in a solar air channel. J Enhanc Heat Transf 21(1):21–49

Kumar CS, Pattamatta A (2015) A numerical study of convective heat transfer enhancement with jet impingement cooling using porous obstacles. J Enhanc Heat Transf 22(4):303–328

Kumar R, Varma HK, Mohanty B, Agrawal KN (2002) Augmentation of heat transfer during filmwise condensation of steam and R-134a over single horizontal finned tubes. Int J Heat Mass Transfer 145(1):201–211

Kumar S, Kothiyal AD, Bisht MS, Kumar A (2019) Effect of nanofluid flow and protrusion ribs on performance in square channels: an experimental investigation. J Enhanc Heat Transf 26 (1):75–100

Kuzenov VV, Ryzhkov SV (2018) Approximate calculation of convective heat transfer near hypersonic aircraft surface. J Enhanc Heat Transf 25(2):181–193

Kuzma-Kichta Y, Leontiev A (2018) Choice and justification of the heat transfer intensification methods. J Enhanc Heat Transf 25(6):465–564

Laohalertdecha S, Wongwises S (2010) The effects of corrugation pitch on the condensation heat transfer coefficient and pressure drop of R-134a inside horizontal corrugated tube. Int J Heat Mass Transf 53:2924–2931

Laohalertdecha S, Naphon P, Wongwises S (2007) A review of electrohydrodynamic enhancement of heat transfer. Renew Sust Energy Rev 11(5):858–876

Lee DH, Song J, Jo MC (2004) The effects of nozzle diameter on impinging jet heat transfer and fluid flow. J Heat Transfer 126:554–557

Lee J, Lee SJ (2000) The effect of nozzle configuration on stagnation region heat transfer enhancement of axisymmetric jet impingement. Int J Heat Mass Transfer 43:3497–3509

Lee SC, Nam SC, Ban TG (1998) Performance of heat transfer and pressure drop in a spirally indented tube. KSME Int J 12(5):917–925

Leontiev AI, Kiselev NA, Burtsev SA, Strongin MM, Vinogradov YA (2016) Experimental investigation of heat transfer and drag on surfaces with spherical dimples. Exp Therm Fluid Sci 79:74–84

Leporini M, Corvaro F, Marchetti B, Polonara F, Benucci M (2018) Experimental and numerical investigation of natural convection in tilted square cavity filled with air. Exp Therm Fluid Sci 99:572–583

Liu T, Cai Z, Lin J (1990) Enhancement of nucleate boiling heat transfer with additives. In: Heat transfer enhancement and energy conservation. CRC Press, Boca Raton, FL

Lopina RF, Bergles AE (1967) Heat transfer and pressure drop in tape generated swirl flow. MIT Dept. of Mechanical Engineering, Cambridge, MA

Lopina RF, Bergles AE (1973) Subcooled boiling of water in tape-generated swirl flow. J Heat Transfer 95(2):281–283

Lou ZQ, Mujumdar AS, Yap C (2005) Effects of geometric parameters on confined impinging jet heat transfer. Appl Therm Eng 25:2687–2697

Luo J, Razinsky EH (2009) Analysis of turbulent flow in 180 deg turning ducts with and without guide vanes. ASME J Turbomach 131(2):021011

Luo L, Wang C, Wang L, Sunden B, Wang S (2015) Computational investigation of dimple effects on heat transfer and friction factor in a Lamilloy cooling structure. J Enhanc Heat Transf 22 (2):147–175

Macbain SM, Bergles AE, Raina S (1997) Heat transfer and pressure drop characteristics of flow boiling in a horizontal deep spirally fluted tube. HVAC&R Res 3(1):65–80

Mahdavi M, Tiari S, De Schampheleire S, Qiu S (2018) Experimental study of the thermal characteristics of a heat pipe. Exp Therm Fluid Sci 93:292–304

Mallor F, Raiola M, Vila CS, Örlü R, Discetti S, Ianiro A (2019) Modal decomposition of flow fields and convective heat transfer maps: an application to wall-proximity square ribs. Exp Therm Fluid Sci 102:517–527

Manglik RM, Bergles AE (2002) Swirl flow heat transfer and pressure drop with twisted-tape inserts. Adv Heat Transfer 36:183–266

Manadilli G (1997) Replace implicit equations with signomial functions. Chem Eng 104(8):129

Manglik RM, Jog MA (2016) Resolving the energy–water nexus in large thermoelectric power plants: a case for application of enhanced heat transfer and high-performance thermal energy storage. J Enhanc Heat Transf 23(4):263–282

Manglik RM, Patel P, Jog MA (2012) Swirl-enhanced laminar forced convection through axially twisted rectangular ducts–part 2, heat transfer. J Enhanc Heat Transf 19(5):437–450

Marco SM, Velkoff HR (1963) Effect of electrostatic fields on free convection heat transfer from flat plates. ASME joint meeting on heat transfer, Boston, MA, ASME paper, 63-HT-9

Martin H (1977) Heat and mass transfer between impinging gas jets and solid surfaces. In: Advances heat transfer, vol 13. Elsevier, Amsterdam, pp 1–60

Matzner B (1965) Critical heat flux in long tubes at 1000psi with and without swirl promoters. ASME-Paper, No. 65-WA-HT-30

Meng H, Zhu G, Yu Y, Wang Z, Wu J (2016) The effect of symmetrical perforated holes on the turbulent heat transfer in the static mixer with modified Kenics segments. Int J Heat Mass Transfer 99:647–659

Metwally HM, Manglik RM (2004) Enhanced heat transfer due to curvature-induced lateral vortices in laminar flows in sinusoidal corrugated-plate channels. Int J Heat Mass Transfer 47 (10–11):2283–2292

Metzger DE, Plevich CW, Fan CS (1984) Pressure loss through sharp 180 deg turns in smooth rectangular channels. J Eng Gas Turbines Power 106(3):677–681

Metzger DE, Sahm MK (1986) Heat transfer around sharp 180 deg turns in smooth rectangular channels. ASME J Heat Transfer 108(3):500–506

Mishra PC, Sen S, Mukhopadhyay A (2014) Experimental investigation of heat transfer characteristics in water shower cooling of steel plate. J Enhanc Heat Transf 21(1):1–20

Mohammed HA, Abuobeida IAA, Vuthaluru HB, Liu S (2019) Two-phase forced convection of nanofluids flow in circular tubes using convergent and divergent conical rings inserts. Int Commun Heat Mass Transf 101:10–20

Moody LF (1944) Friction factors for pipe flows. Trans ASME 66:671–684

Moon MA, Kim KY (2013) Computational analysis of trailing edge internal cooling of a gas turbine blade with pin-fin arrays. J Enhanc Heat Transf 20(2):137–151

Morgan AI, Bromley LA, Wilke CR (1949) Effect of surface tension on heat transfer in boiling. Ind Eng Chem 41(12):2767–2769

Motamed Ekitesabi M, Sako M, Chiba T (1987) Fluid flow and heat transfer in wavy sinusoidal channels: 1st report, numerical analysis of two dimensional laminar flow field: series b: fluid engineering, heat transfer, combustion, power, thermophysical properties. JSME Int Journal Bull JSME 30(269):1854

Muley A, Manglik RM (2000) Enhanced thermal-hydraulic performance optimization of chevron plate heat exchangers. Int J Heat Exchangers 1(1):3–18

Murase T, Wang HS, Rose JW (2006) Effect of inundation for condensation of steam on smooth and enhanced condenser tubes. Int J Heat Mass Transfer 49(17–18):3180–3189

Naik MT, Fahad SS, Sundar LS, Singh MK (2014) Comparative study on thermal performance of twisted tape and wire coil inserts in turbulent flow using CuO/water nanofluid. Exp Therm Fluid Sci 57:65–76

Naik B, Vinod AV (2018) Heat transfer enhancement using non-Newtonian nanofluids in a shell and helical coil heat exchanger. Exp Therm Fluid Sci 90:132–142

Naik H, Tiwari S (2018) Effect of aspect ratio and arrangement of surface-mounted circular cylinders on heat transfer characteristics. J Enhanc Heat Transf 25(4–5):443–463

Naresh Y, Vignesh KS, Balaji C (2018) Experimental investigations of the thermal performance of self-rewetting fluids in internally finned wickless heat pipes. Exp Therm Fluid Sci 92:436–446

Nasr A, Debbissi C, Orfi J, Nassrallah S (2009) Evaporation of water by natural convection in partially wetted heated vertical plates: effect of the number of the wetted zone. J Eng Appl Sci 4 (1):51–59

Nishimura T, Kajimoto Y, Kawamura Y (1986) Mass transfer enhancement in channels with a wavy wall. J Chem Eng Jpn 19(2):142–144

Nishimura T, Yano K, Yoshino T, Kawamura Y (1990) Occurrence and structure of Taylor–Goertler vortices induced in two-dimensional wavy channels for steady flow. J Chem Eng Jpn 23(6):697–703

Noh SW, Suh KY (2014) Critical heat flux in various inclined rectangular straight surface channels. Exp Therm Fluid Sci 52:1–11

Nouri NM, Sarreshtehdari A (2009) An experimental study on the effect of air bubble injection on the flow induced rotational hub. Exp Therm Fluid Sci 33(2):386–392

Nouri-Borujerdi A, Nakhchi ME (2018) Experimental study of convective heat transfer in the entrance region of an annulus with an external grooved surface. Exp Therm Fluid Sci 98:557–562

Nozu S, Honda L, Nakata H (1995) Condensation of refrigerants CFCLL and CFCLL3 in the annulus of a double-tube coil with an enhanced inner tube. Exp Thermal Fluid Sci 11:40–51

O'Brien JE, Sparrow EM (1982) Corrugated-duct heat transfer, pressure drop, and flow visualization. J Heat Transfer 104(3):410–416

O'Donovan TS, Murray DB (2007) Jet impingement heat transfer, part I: mean and root mean-square heat transfer and velocity distributions. Int J Heat Mass Transfer 50:3291–3301

Ohadi M, Darabi J, Roget B (2000) Electrode design, fabrication, and materials science for EHD-enhanced heat and mass transport. Annu Rev Heat Transfer 11(11):563–632

Ökten K, Biyikoglu A (2018) Effect of air bubble injection on the overall heat transfer coefficient. J Enhanc Heat Transf 25(3):195–210

Pal PK, Saha SK (2014) Experimental investigation of laminar flow of viscous oil through a circular tube having integral spiral corrugation roughness and fitted with twisted tapes with oblique teeth. Exp Thermal Fluid Sci 57:301–309

Parsons JA, Han JC, Zhang YM (2002) Effects of model orientation and wall heating condition on local heat transfer in a rotating two-pass square channel with rib turbulators. Int J Heat Mass Transfer 38(7):1151–1159

Patil RH (2017) Experimental studies on heat transfer to Newtonian fluids through spiral coils. Exp Therm Fluid Sci 84:144–155

Perng SW, Wu HW (2013) Heat transfer enhancement for turbulent mixed convection in reciprocating channels by various rib installations. J Enhanc Heat Transf 20(2):95–114

Petukhov BS (1970) Heat transfer and friction in turbulent pipe flow with variable physical properties. In: Advances in heat transfer, vol 6. Elsevier, Amsterdam, pp 503–564

Piasecka M, Maciejewska B (2015) Heat transfer coefficient during flow boiling in a minichannel at variable spatial orientation. Exp Therm Fluid Sci 68:459–467

Podsushnyy AM, Minayev AN, Statsenko VN, Yakubovskiy YV (1980) Effect of surfactants and of scale formation on boiling heat transfer to sea water. Heat Transfer Sov Res 12:113–114

Prajapati YK, Pathak M, Khan MK (2016) Transient heat transfer characteristics of segmented finned microchannels. Exp Therm Fluid Sci 79:134–142

Qu J, Li X, Wang Q, Liu F, Guo H (2017) Heat transfer characteristics of micro-grooved oscillating heat pipes. Exp Therm Fluid Sci 85:75–84

Rabas TJ, Webb RL, Thors P, Kim NK (2017) Performance of three-dimensional helically dimpled tubes influenced by roughness shape and spacing. J Enhanc Heat Transf 24(1–6):117–128

Rabas TJ, Webb RL, Thors P, Kim NK (1994) Influence of roughness shape and spacing on the performance of three-dimensional helically dimpled tubes. J Enhanc Heat Transf 1(1)

Rabas TJ, Thors P, Webb RL, Kim N-H (1993) Influence of roughness shape and spacing on the performance of three-dimensional helically dimpled tubes. J Enhanc Heat Transf 1:53–64

Rao DVR, Babu CS, Prabhu SV (2004) Effect of turn region treatments on the pressure loss distribution in a smooth square channel with sharp 180 bend. Int J Rotating Mach 10(6):459–468

Rao DVR, Prabhu SV (2004) Pressure drop distribution in smooth and rib roughened square channel with sharp 180 bend in the presence of guide vanes. Int J Rotating Mach 10(2):99–114

Rainieri S, Farina A, Pagliarini G (1996) Experimental investigation of heat transfer and pressure drop augmentation for laminar flow in spirally enhanced tubes. In: Pisa ETS (ed) Proceedings of the 2nd European thermal-sciences and 14th UIT National heat transfer conference, vol 1, Roma, pp 203–209

Rainieri S, Pagliarini G (1997) Convective heat transfer to orange juice in smooth and corrugated tubes. Int J Heat Technol 15(2):69–75

Ravigururajan TS, Bergles AE (1996) General correlations for pressure drop and heat transfer for single-phase turbulent flow in internally ribbed tubes. Exp Thermal Fluid Sci 13:55–70

Reid RS (1986) Augmented in-tube evaporation of refrigerant 113. Master of science thesis, Iowa State University, Ames, Iowa

Rout PK, Saha SK (2013) Laminar flow heat transfer and pressure drop in a circular tube having wire-coil and helical screw-tape inserts. J Heat Transfer 135(2):021901

Rush TA, Newell TA, Jacobi AM (1999) An experimental study of flow and heat transfer in sinusoidal wavy passages. Int J Heat Mass Transfer 42(9):1541–1553

Saha SK (2010) Thermohydraulics of laminar flow through rectangular and square ducts with axial corrugation roughness and twisted tapes with oblique teeth. J Heat Transf 132:081701

Saha SK, Swain BN, Dayanidhi B (2012) Friction and thermal characteristics of laminar flow of viscous oil through a circular tube having axial corrugations and fitted with helical screw-tape inserts. J Fluids Eng 134:051210

Saha S, Saha SK (2013) Enhancement of heat transfer of laminar flow of viscous oil through a circular tube having integral helical rib roughness and fitted with helical screw-tapes. Exp Therm Fluid Sci 47:81–89

Saha G, Paul MC (2018) Investigation of the characteristics of nanofluids flow and heat transfer in a pipe using a single phase model. Int Commun Heat Mass Transf 93:48–59

Sahebi M, Alemrajabi AA (2014) Electrohydrodynamic (EHD) enhancement of natural convection heat transfer from a heated inclined plate. J Enhanc Heat Transf 21(1):51–61

Saini RP, Saini JS (1997) Heat transfer and friction factor correlations for artificially roughened ducts with expanded metal mesh as roughened element. Int J Heat Mass Transfer 40:973–986

Salim MM, France DM, Panchal CB (1999) Heat transfer enhancement on outer surface of spirally indented tubes. J Enhanc Heat Transf 6:327–341

Saltanov GA, Kukushkin AN, Solodov AP, Sotskov SA, Jakusheva EV, Chempik E (1986) Surfactant influence on heat transfer at boiling and condensation, In: Heat transfer, Hemisphere, Washington, 5, 2245–2250

San JY, Huang WC, Chen CA (2015) Experimental investigation on heat transfer and fluid friction correlations for circular tubes with coiled-wire inserts. Int Commun Heat Mass Transfer 65:8–14

Sandhu H, Gangacharyulu D, Singh MK (2018) Experimental investigations on the cooling performance of microchannels using alumina nanofluids with different base fluids. J Enhanc Heat Transf 25(3):283–291

Sara BA, Bali T (2007) An experimental study on heat transfer and pressure drop characteristics of decaying swirl flow through a circular pipe with a vortex generator. Exp Therm Fluid Sci 32 (1):158–165

Saysroy A, Eiamsa-ard S (2017) Periodically fully-developed heat and fluid flow behaviors in a turbulent tube flow with square-cut twisted tape inserts. Appl Therm Eng 112:895–910

Schabacker J, Boelcs A, Johnson BV (1998) PIV investigation of the flow characteristics in an internal coolant passage with two ducts connected by a sharp 180 deg bend. In: Proc ASME turbo expo conf paper no. 98-GT-544

Sephton HH (1971) Interface enhancement for vertical tube evaporators-novel way of substantially augmenting heat and mass transfer. In: Mechanical engineering, vol 93. ASME, New York, p 1157

Shafaee M, Alimardani F, Mohseni SG (2016) An empirical study on evaporation heat transfer characteristics and flow pattern visualization in tubes with coiled wire inserts. Int Commun Heat Mass Transfer 76:301–307

Sharma N, Tariq A, Mishra M (2018) Detailed heat transfer and fluid flow investigation in a rectangular duct with truncated prismatic ribs. Exp Therm Fluid Sci 96:383–396

Shatto DP, Peterson GP (2017) Flow boiling heat transfer with twisted tape inserts. J Enhanc Heat Transf 24(1–6):21–46

Siddique M, Alhazmy M (2008) Experimental study of turbulent single-phase flow and heat transfer inside a micro-finned tube. Int J Refrig 31(2):234–241

Sohel MR, Khaleduzzaman SS, Saidur R, Hepbasli A, Sabri MFM, Mahbubul IM (2014) An experimental investigation of heat transfer enhancement of a minichannel heat sink using Al_2O_3–H_2O nanofluid. Int J Heat Mass Transfer 74:164–172

Son SY, Kihm KD, Han JC (2002) PIV flow measurements for heat transfer characterization in two-pass square channels with smooth and 90 ribbed walls. Int J Heat Mass Transfer 45(24):4809–4822

Song K, Xi Z, Su M, Wang L, Wu X, Wang L (2017) Effect of geometric size of curved delta winglet vortex generators and tube pitch on heat transfer characteristics of fin-tube heat exchanger. Exp Therm Fluid Sci 82:8–18

Sonnad JR, Goudar CT (2006) Turbulent flow friction factor calculation using a mathematically exact alternative to the Colebrook–White equation. J Hydraul Eng 132(8):863–867

Steenbergen W, Voskamp J (1998) The rate of decay of swirl in turbulent pipe flow. Flow Meas Instrum 9(2):67–78

Stevens J, Webb BW (1991) Local heat transfer coefficients under an axisymmetric, single-phase liquid jet. J Heat Transfer 113:71–78

Stevens J, Webb BW (1992) Measurements of the free surface flow structure under an impinging, free liquid jet. J Heat Transfer 114:79–84

Suga K, Aoki H (2017) Heat transfer and pressure drop in multilouvered fins. J Enhanc Heat Transf 24(1–6):137–144

Swamee PK, Jain AK (1976) Explicit equations for pipe-flow problems. J Hydraul Div ASCE 102(5):657–664

Tada Y, Takimoto A, Hayashi Y (1997) Heat transfer enhancement in a convective field by applying ionic wind. J Enhanc Heat Transf 4(2):71–86

Tarasov GI, Shchukin VK (1977) An experimental study of heat transfer in channels equipped with extended screw-type intensifiers. In: Heat-and-mass transfer in aircraft engines, vol 1. Kazan Aviation Institute, Kazan, Russia, pp 40–45

Taslim ME, Li T, Krecher DM (1996) Experimental heat transfer and friction in channel roughened with angled, v-shaped and discrete ribs on two opposite walls. Trans ASME J Turbomach 118:20–28

Taylor RP, Hodge BK (2017) A review of fully-developed Nusselt numbers and friction factors in pipes with 3-dimensional roughness. J Enhanc Heat Transf 24(1–6):357–370

Terekhov V, Khafaji H, Ekaid A (2015) Numerical simulation for laminar forced convection in a horizontal Insulated Channel with wetted walls, Proc 8th ICCHMT, Istanbul, May 25–28, 2015

Terekhov VI, Khafaji HQ, Gorbachev MV (2018) Heat and mass transfer enhancement in laminar forced convection wet channel flows with uniform wall heat flux. J Enhanc Heat Transf 25(6):565–577

Thonon B, Vidil R, Marvillet C (2017) Plate heat exchangers: research and developments. J Enhanc Heat Transf 24(1–6):129–136

Tzan YL, Yang YM (1990) Experimental study of surfactant effects on pool boiling heat transfer. J Heat Transfer 112(1):207–212

Uddin N, Weigand B, Younis BA (2019) Comparative study on heat transfer enhancement by turbulent impinging jet under conditions of swirl, active excitations and passive excitations. Int Commun Heat Mass Transfer 100:35–41

Ustimenko BP (1977) Processes of turbulent carrying over in twirled currents. Nauka, Alma-Ata, USSR

Vulchanov NL, Zimparov VD, Delov LB (1991) Heat transfer and friction characteristics of spirally corrugated tubes for power plant condensers – 2. A mixing-length model for predicting fluid friction and heat transfer. Int J Heat Mass Transf 34(9):2199–2206

Varun, Garg MO, Nautiyal H, Khurana S, Shukla MK (2016) Heat transfer augmentation using twisted tape inserts. A review. Renew Sust Energy Rev 63:193–225

Vasiliev LL, Zhuravlyov AS, Shapovalov A (2012) Hear transfer enhancement in mini channels with micro/nano particles deposited on a heat-loaded wall. J Enhanc Heat Transf 19:3–24

Vilemas Y, Poshkas P (1992) Heat transfer in gas-cooled channels under the effect of thermal-gravity and centrifugal forces. Academia, Vilnius. (in Russian)

Viskanta R (1961) Critical heat flux for water in swirling flow. Nucl Sci Eng 10(2):202–203

Vyas S, Manglik RM, Jog MA (2010) Visualization and characterization of a lateral swirl flow structure in sinusoidal corrugated-plate channels. J Flow Vis Image Process 17(4):281–296

Volchkov E, Terekhov VV, Terekhov VI (2004) A numerical study of boundary layer heat and mass transfer in a forced convection of humid air with surface steam condensation. Int J Heat Mass Transf 47:1473–1481

Wang CC (2017) Fin-and-tube heat exchangers: recent patents. J Enhanc Heat Transf 24 (1–6):255–268

Wang CC, Chiou CB, Lu DC (1996) Single-phase heat transfer and flow friction correlations for microfin tubes. Int J Heat Fluid Flow 17:500–508

Wang L, Sun D, Liang P, Zhuang L, Tan Y (2000) Heat transfer characteristics of carbon steel spirally fluted tube for high pressure preheaters. Energy Convers Manag 41:993–1005

Wang TAA, Hartnett JP (1994) Pool boiling heat transfer from a horizontal wire to aqueous surfactant solutions. Heat Transf., I Chem. E, UK 5:177–182

Warrier G, Dhir DV (2013) Comparison of heat removal using miniature channels, jets, and sprays. J Enhanc Heat Transf 20(1):17–32

Wasekar VM, Manglik RM (2018) Enhanced heat transfer in nucleate pool boiling of aqueous surfactant and polymeric solutions. J Enhanc Heat Transf 24(1–6):47–62

Webb BW, Ma CF (1995) Single-phase liquid jet impingement heat transfer. Adv Heat Transfer 26:105–217

Webb RL, Chamra LM (1991) On-line cleaning of particulate fouling in enhanced tubes. In: Fouling and enhancement interactions. ASME, pp 47–54

Webb RL, Kim NH (2005) Principles of enhanced heat transfer. Taylor & Francis, Boca Raton, FL

Wei-Mon Y (1992) Effects of film evaporation on laminar mixed convection heat and mass transfer in a vertical channel. Int J Heat Mass Transf 12:3419–3429

Wen MY, Jang KJ, Ho CY (2015) Flow boiling heat transfer in R-600a flows inside an annular tube with metallic porous inserts. J Enhanc Heat Transf 22(1):47–65

Wijayanta AT, Istanto T, Kariya K, Miyara A (2017) Heat transfer enhancement of internal flow by inserting punched delta winglet vortex generators with various attack angles. Exp Therm Fluid Sci 87:141–148

Wu WT, Yang YM (1992) Enhanced boiling heat transfer by surfactant additives. In: Pool and external flow boiling. ASME, New York, pp 361–366

Yan W, Lin T (1988) Combined heat and mass transfer in laminar forced Convection Channel flows. Int Commun Heat Mass Transf 15:333–343

Yakovlev AB (2013) Heat transfer and hydraulic resistance in single-phase forced convection in annular channels with twisting wire inserts. J Enhanc Heat Transf 20(6):519–525

Yang H, Wen J, Tong X, Li K, Wang S, Li Y (2016) Numerical investigation on configuration improvement of a plate-fin heat exchanger with perforated wing-panel header. J Enhanc Heat Transf 23(1):1–21

Yang LC, Asako Y, Yamaguchi Y, Faghri M (1997) Numerical prediction of transitional characteristics of flow and heat transfer in a corrugated duct. J Heat Transfer 119(1):62–69

Yang YM, Maa JR (1983) Pool boiling of dilute surfactant solutions. J Heat Transfer 105 (1):190–192

Yerra KK, Manglik RM, Jog MA (2007) Optimization of heat transfer enhancement in single-phase tube-side flows with twisted-tape inserts. Int J Heat Exchangers 8(1):117

Yilmaz M, Comakli O, Yapici S, Sara ON (2005) Performance evaluation criteria for heat exchangers based on first law analysis. J Enhanc Heat Transf 12(2):121–158

Yilmaz M, Sara ON, Karsli S (2001) Performance evaluation criteria for heat exchangers based on second law analysis. Exergy Int J 1(4):278–294

Yilmaz MC, omakli O, Yapici S, Sara ON (2003) Heat transfer and friction characteristics in decaying swirl flow generated by different radial guide vane swirl generators. Energy Convers Manag 44(2):283–300

Yonggang Y, Junping H, Zhongliang AI, Lanjun Y, Qiaogen Z (2006) Experimental studies of the enhanced heat transfer from a heating vertical flat plate by ionic wind. Plasma Sci Technol 8 (6):697

Zdaniuk GJ, Chamra LM, Mago PJ (2008) Experimental determination of heat transfer and friction in helically-finned tubes. Exp Therm Fluid Sci 32(3):761–775

Zhai Y, Xia G, Li Z, Wang H (2017) Experimental investigation and empirical correlations of single and laminar convective heat transfer in microchannel heat sinks. Exp Therm Fluid Sci 83:207–214

Zhang B, Wang Y, Zhang J, Li Q (2017a) Experimental research on pressure drop fluctuation of two-phase flow in single horizontal mini-channels. Exp Therm and Fluid Sci 88:160–170

Zhang C, Wang D, Zhu Y, Han Y, Wu J, Peng X (2015) Numerical study on heat transfer and flow characteristics of a tube fitted with double spiral spring. Int J Therm Sci 94:18–27

Zhang J, Diao Y, Zhao Y, Zhang Y (2017b) An experimental investigation of heat transfer enhancement in minichannel: combination of nanofluid and micro fin structure techniques. Exp Therm Fluid Sci 81:21–32

Zhang J, Kundu J, Manglik RM (2004) Effect of fin waviness and spacing on the lateral vortex structure and laminar heat transfer in wavy-plate-fin cores. Int J Heat Mass Transfer 47 (8–9):1719–1730

Zhang L, Yang S, Xu H (2012) Experimental study on condensation heat transfer characteristics of steam on horizontal twisted elliptical tubes. Appl Energy 97:881–887

Zhang YM, Han JC, Lee CP (2017c) Turbulent flow in circular tubes with twisted-tape inserts and axial interrupted ribs. J Enhanc Heat Transf 24(1–6):243–254

Zhou J, Luo X, Feng Z, Xiao J, Zhang J, Guo F, Li H (2017) Saturated flow boiling heat transfer investigation for nanofluid in minichannel. Exp Therm Fluid Sci 85:189–200

Zimparov VD, Vulchanov NL, Delov LB (1991) Heat transfer and friction characteristics of spirally corrugated tubes for power plant condensers – 1. Experimental investigation and performance evaluation. Int J Heat Mass Transf 34(9):2187–2197

Zimparov VD, Bonev PJ, Petkov VM (2016) Benefits from the use of enhanced heat transfer surfaces in heat exchanger design: a critical review of performance evaluation. J Enhanc Heat Transf 23(5)

Chapter 3
Heat Exchanger Design Theory, Fin Efficiency, Variation of Fluid Properties

Heat transfer enhancement techniques are developed with the understanding of heat transfer fundamental and heat exchanger design theory. The idea of heat transfer enhancement may be applied and exploited to any type of heat exchanger or heat exchange device. Heat exchanger design calculations deal with rating problem and sizing problem. The heat exchanger rating involves the consideration of heat exchanger type, size and surface geometry. The process conditions include flow rate, entering fluid condition and fouling factor. Heat transfer rate and pressure drop of the heat-exchanging fluids are calculated. The heat exchanger sizing problem involves the consideration of process requirements like flow rate, entering fluid conditions and acceptable pressure drops.

Table 3.1 shows heat exchanger design considerations. Design selections are made from the consideration of design specifications. Decision has to be made with regard to heat exchanger materials, heat exchanger type and flow arrangement and heat transfer surface geometries. Thermal and hydraulic analysis was done. Optimum design necessitates the consideration of fluid velocities and heat transfer surface geometries. The optimisation may be from the point of view of first cost, operating cost, life cycle cost and size dimensions. The optimisation process needs parametric analysis to understand the trade-offs among the possible design variables (Shah et al. 1978). The fundamentals of heat exchanger design theory and the overall thermal resistance from the consideration of individual thermal resistances may be obtained from any basic heat transfer text. If the temperatures at four cardinal points of two heat-exchanging fluids are known, the LMTD concept may be used. If any of the four cardinal point temperature is unknown, then the number of transfer unit (NTU) concept may be utilised.

The relative arrangement of two flowing streams in the heat exchanger may be parallel, counterflow or crossflow configurations. The geometry correction factor "F" is discussed in Webb and DiGiovanni (1989) and it provides a method to predict the performance for partially mixed conditions. The LMTD design method and NTU

Table 3.1 Heat exchanger design considerations (Webb and Kim 2005)

Design specifications	Design selection
Process requirements	*Heat exchanger materials*
(a) Fluid compositions, and inlet flow conditions (flow rate, temperature and pressure) (b) Heat duty or required exit temperatures (c) Allowable pressure drops	(a) Fluid temperatures and pressures (b) Corrosive characteristics of fluid-material combination
Operating and maintenance considerations	*Heat exchanger type*
(a) Fouling potential and method of cleaning (b) Failure due to corrosion, thermal stress, vibration or freezing (c) Repair of leaks (d) Part load operating characteristics	(a) Design pressure and fluid temperatures (b) Corrosion, stress, vibration and freezing considerations (c) Fouling potential and cleaning possibilities (d) First cost, operating and maintenance costs
Size and weight restrictions	*Heat transfer surface geometries*
(a) Frontal area, length or height (b) Possible weight restrictions	(a) Thermal resistance (b) Potential for use of enhanced surfaces (c) Fouling potential and cleaning possibilities (d) Unit cost of heat transfer surface

design method differ only in the algebraic form of the resulting equations, but essentially the two design methods are equivalent. The NTU method gives us the opportunity of physical interpretation of the thermodynamic performance of the heat exchanger, which is not provided by LMTD method. Readily available algebraic form in ε-NTU method is important for digital computer.

The design of shell and tube heat exchangers for enhanced heat transfer rates is an arduous job. This is due to lack of proper correlations and expressions to calculate the log mean temperature difference (LMTD) correction factor. Though there are different charts available to determine the LMTD correction factor, the steep regions of the chart pose difficulty in collecting data. Figure 3.1 shows the charts used for calculating the LMTD correction factor for one shell pass and even number of tube passes. Thus, Fakheri (2017) proposed a single correlation to calculate the LMTD correction factor for shell and tube heat exchangers with any number of shells and even number of tube passes.

3.1 Fin Efficiency

Fin is the extended surface and it provides reduced thermal resistance for the flow stream. The heat conductance of a finned surface must be multiplied by fin efficiency factor to take the temperature gradient in the fin into account. The fin efficiency, η_f, is

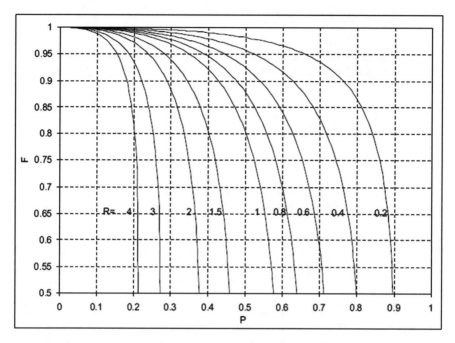

Fig. 3.1 Chart for calculating the LMTD correction factor (Fakheri 2017)

the ratio of the actual heat transfer from the fin to that which would occur if the entire fin were at its base temperature. A finned surface heat exchanger consists of the secondary finned surface and the primary surface to which the fins are attached. Total surface efficiency accounts for the efficiency of the composite structure consisting of the fins and the base surface. The total surface efficiency η is given by

$$\eta = 1 - (1 - \eta_f)^{A_f}/_A \tag{3.1}$$

where η_f is the fin efficiency, A_f is the area of the extended surface and A is the heat transfer surface area.

3.2   Heat Transfer Coefficients and Friction Factor

The recommended design equation and correlation for smooth surfaces in single-phase forced convective heat transfer are presented below. In the periphery of tube either inside or outside, heat transfer coefficient and equations for heat exchangers are presented sequentially considering the parameters like laminar flow, turbulent flow, duct and flat plate.

3.2.1   Laminar Flow in Ducts

The solution and equation have been given by Blasius. The Blasius solution for a plate of length L for the average Stanton number with laminar flow is given as

$$St = 0.664(Re_L)^{-0.5}(Pr)^{-\frac{2}{3}} \tag{3.2}$$

And friction factor is given as

$$\bar{f} = 1.328 Re_L^{-0} \tag{3.3}$$

Table 3.2 shows fully developed laminar flow solution. The table shows the friction factor and Nusselt number for different geometries. Also, the value of

Table 3.2 Solutions for fully developed laminar flow (Webb and Kim 2005)

Geometry	Nu_H	Nu_r	fRe	K_∞	j_H/f	j_r/f	L_{hy}^+
$\dfrac{2b}{2a} = 0$	8.235	7.541	24.00	0.686	0.386	0.354	0.0056
$\dfrac{2b}{2a} = \dfrac{1}{8}$	6.490	5.597	20.585	0.879	0.355	0.306	0.0094
$\dfrac{2b}{2a} = \dfrac{1}{6}$	6.049	5.137	19.702	0.945	0.346	0.294	0.0010
$\dfrac{2b}{2a} = \dfrac{1}{4}$	5.331	4.439	18.233	1.076	0.329	0.274	0.0147
◯	4.364	3.657	16.00	1.24	0.307	0.258	0.038
$\dfrac{2b}{2a} = \dfrac{1}{2}$	4.123	3.391	15.548	1.383	0.299	0.245	0.0255
$\dfrac{2b}{2a} = 1$	3.608	3.091	14.227	1.552	0.286	0.236	0.0324
$\dfrac{2a}{2b} = \dfrac{\sqrt{3}}{2}$	3.111	2.47	13.333	1.818	0.263	0.209	0.0398
$\dfrac{2b}{2a} = \dfrac{\sqrt{3}}{2}$	3.014	2.39	12.630	1.739	0.269	0.214	0.0408
$\dfrac{2b}{2a} = 2$	2.880	2.22	13.026	1.991	0.249	0.192	0.0443
$\dfrac{2b}{2a} = .25$	2.600	1.99	12.622	2.236	0.232	0.178	0.0515
$j/f \equiv \dfrac{NuPr^{-1/3}}{fRe}$	j_H, j_T for $Pr = 0.7$						

H constant heat flux boundary condition, T constant temperature boundary condition

Nusselt number is arranged in decreasing order. It can be easily recognised that the shape and geometry affect Nusselt number. The table shows the ratio of $j/_f$ for other geometries like circular tubes and triangular channels.

3.2.2   Turbulent Flow in Ducts

The most relevant and precise equation for heat transfer in smooth tubes is as follows:

$$St = \frac{f/2}{1.07 + 12.7(f/2)^{\frac{1}{2}}\left(Pr^{\frac{2}{3}} - 1\right)} \tag{3.4}$$

And the friction factor

$$f = (1.58 \ln Re_d - 3.28)^{-2} \tag{3.5}$$

3.2.3   Tube Banks (Single-Phase Flow)

The equations commonly used are

$$\overline{St} = 0.664 Re_L^{-0.5} Pr^{-\frac{2}{3}} \tag{3.6}$$

And average friction factor is

$$\bar{f} = 1.328 Re_L^{-0.5} \tag{3.7}$$

3.2.4   Film Condensation and Nucleate Boiling

All equations had considered heat transfer coefficient as the average value over length of plate or diameter of tube.

1. Gravity-drained laminar film average condensation coefficient on a vertical plate:

$$\bar{h} = 0.943 \left(\frac{K^3 g(\rho_l - \rho_v)\lambda}{v_l \Delta T_{vs} L}\right)^{\frac{1}{4}} \tag{3.8}$$

2. For inclined plane, gravity force is $g \sin\theta$. Thus equation for heat transfer coefficient is

$$\bar{h} = 0.943 \left(\frac{K^3 g \sin\theta (\rho_1 - \rho_v{}^v)\lambda}{V_1} \right)^{1/4} \tag{3.9}$$

It can be transformed into condensate Reynolds number form. Condensate Reynolds number is

$$Re_L = 4\frac{\Gamma}{\mu} \tag{3.10}$$

where Γ is the condensate flow rate per unit plate width.
Thus, condensation coefficient is

$$\bar{h} = 1.47 \left(\frac{K^3 \rho_1 (\rho_1 - \rho_v)g}{\mu_1^2} \right) Re_L^{-\frac{1}{3}} \tag{3.11}$$

3. For horizontal tube,
   The condensation coefficient for laminar film is

$$\bar{h} = 0.728 \left(\frac{K_g^3 (\rho_1 - \rho_v)\lambda}{v_1 \Delta T_{vs} d} \right)^{\frac{1}{4}} \tag{3.12}$$

The transformed Reynolds number draining from the tube is

$$\bar{h} = 1.51 \left(\frac{K^3 \rho_1 (\rho_1 - \rho_v)g}{\mu_1^2} \right)^{1/3} Re_L^{-1} \tag{3.13}$$

4. The condensation coefficient for local turbulent film on a vertical plate is

$$h = 0.0038 \left(\frac{K^3 \rho_1 (\rho_1 - \rho_v)g}{\mu_1^2} \right)^{\frac{1}{3}} Re_L^{0.4} Pr^{2/3} \tag{3.14}$$

5. Nucleate Boiling
   The precise correlation for horizontal plain tube nucleate boiling is

$$h = 90 q^{2/3} M^{1/2} p_r^m (-\log_{10} p_r)^{-0.55} \tag{3.15}$$

where $m = 0.12 - 0.2 \log_{10} R_p$ and R_p is roughness of surface (in μm).

3.3 Effect of Fluid Properties

Fluid properties depend on the fluid temperature and the heat transfer coefficient also gets affected. Temperature is also sensitive to the length of the entrance region. The entrance region in turn also depends on the viscosity of the fluids like polymers and oils. The accurate design of the heat exchanger needs to perform incremental design which is possible only when well-adapted computer methods are applied. The effect of variation of fluid properties may be taken care of by correcting heat transfer and friction correlations for fluid property variation across the boundary layer. This is done either by evaluating the properties at the film temperature which is the average of local mixed fluid temperature and the wall temperature or by a fluid property ratio evaluated at the mixed fluid and the wall temperature. Correction of fluid properties is given in Tables 3.3 and 3.4. Again, fluid property correction may be done by using graphical plots as shown in Figs. 3.2 and 3.3.

Table 3.3 Correction for fluid properties (laminar flow) (Kays and London 1984)

	Heating	Cooling
Liquids		
$\dfrac{St}{St_{cp}} = \left(\dfrac{\mu}{\mu_w}\right)^n$	$n = 0.14$	$n = 0.14$
$\dfrac{f}{f_{cp}} = \left(\dfrac{\mu}{\mu_w}\right)^m$	$m = -0.58$	$m = -0.50$
Gases		
$\dfrac{St}{St_{cp}} = \left(\dfrac{T}{T_w}\right)^n$	$n = 0.0$	$n = 0.0$
$f(T)^m$	$m = -1.0$	$m = -1.0$

Table 3.4 Correction for fluid properties (turbulent flow) (Kays and London 1984)

	Heating	Cooling
Liquids		
$\dfrac{St}{St_{cp}} = \left(\dfrac{\mu}{\mu_w}\right)^n$	$n = 0.11$	$n = 0.25$
$\dfrac{f}{f_{cp}} = \left(\dfrac{\mu}{\mu_w}\right)^m$	$m = \dfrac{1}{6}\left(7 - \dfrac{\mu}{\mu_w}\right)$ $m = -0.25$	$m = -0.24$
Gases		
$\dfrac{St}{St_{cp}} = \left(\dfrac{T}{T_w}\right)^n$	$n = 0.3\log_{10}\dfrac{T_w}{T} + 0.36$ $n = -0.5$	$n = 0.36$
$\dfrac{f}{f_{cp}} = \left(\dfrac{T}{T_w}\right)^m$	$m = 0.6 - 56\left(Re\dfrac{\rho_w}{\rho}\right)^{-0.38}$ $m = -0.1$	$m = 0.6 - 7.9\left(Re\dfrac{\rho_w}{\rho}\right)^{-0.11}$ $m = -0.1$

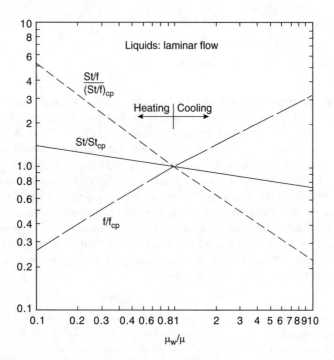

Fig. 3.2 Fluid property correction variation vs. μ_w/μ for laminar flow of liquids (Kays and London 1964)

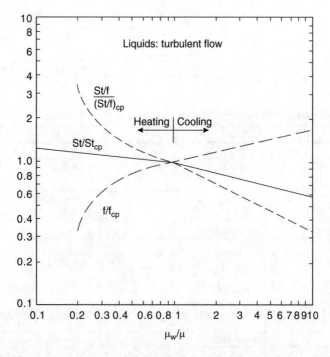

Fig. 3.3 Fluid property correction variation vs. μ/μ_s for turbulent flow of gases (Webb and Kim 2005)

3.4 Reynolds Analogy

Reynolds analogy can be used for the evaluation of thermal data from the data for momentum loss and pressure distribution. This analogy due to Colburn and du Pont de (1933) may be applied for both laminar and turbulent flow. The analogy is mathematically stated as

$$St\,Pr^{\frac{2}{3}} \equiv j = \frac{f}{2} \qquad (3.16)$$

This analogy of thermal energy transport and the hydrodynamic momentum transport for plain surfaces may be equally applied for enhanced surfaces.

References

Colburn AP, du Pont de EI (1933) Mean temperature difference and heat transfer coefficient in liquid heat exchangers. Ind Eng Chem 25(8):873–877

Fakheri A (2017) Determining log mean temperature difference correction factor and number of shells of Shell and tube heat exchangers. J Enhanc Heat Transf 24(1–6):291–304

Kays WM, London AL (1964) Compact heat exchangers. McGraw-Hill, New York

Kays WM, London AL (1984) Compact heat exchangers, 3rd ed., McGraw-Hill, New York

Shah RK, Afimiwala KA, Mayne RW (1978) Heat exchanger optimization. In: Proceedings of 6th Int. heat transfer conference, vol 4, pp 185–191

Webb RL, DiGiovanni MA (1989) Uncertainty in effectiveness-NTU calculations for crossflow heat exchangers. Heat Transfer Eng 10(3):61–70

Webb RL, Kim NY (2005) Principles of enhanced heat transfer. Taylor and Francis, NY

Chapter 4
Fouling on Various Types of Enhanced Heat Transfer Surfaces

Fouling on heat-exchanging surfaces due to deposits on the surface is a very common phenomenon. Fouling essentially increases the thermal impedance and this factor is taken care of by overdesigning the heat-exchanging surfaces. The fouling factor, F_f, is given by

$$F_f = \frac{1}{U_{fouled}} - \frac{1}{U_{clean}} \qquad (4.1)$$

The fouling deposit on the heat-exchanging surface increases the gas pressure drop. This results in a lower gas flow rate and consequently the heat transfer rate is also reduced. The heat-exchanging surfaces are often taken out of service for cleaning of fouling deposits. This cleaning may be of various types: mechanical, chemical or thermal cleaning. The details of fouling may be obtained from Leitner (1980). It is very difficult to make any qualitative prediction of the fouling resistances that will occur on smooth or enhanced surfaces in actual applications. Somerscales and Kundsen (1979), Watkinson (1990, 1991), Bergles and Somerscales (1995), Garrett-Price et al. (1985), Melo et al. (1987) and Bott (1995) provide reviews of fouling on enhanced heat transfer surfaces: Different types of fouling for liquids and gases are

- Precipitation fouling (scaling) with salts precipitating on a heat transfer surface, if the temperature makes supersaturation of the salts at the surface temperature
- Particulate fouling taking place when suspended solids deposit on the surface
- Corrosion fouling when the heat transfer surface material reacts with the fluid
- Chemical reaction at the heat transfer surface that may itself cause surface deposits
- Biofouling deposits when biological mechanism attaches and grows on the heat transfer surface solidification fouling with liquids or gases

4.1   Fouling Fundamentals: Gases and Liquids

Epstein (1983a, b and 1988a, b) describes fouling mechanisms and gives models to predict the deposition and removal rates. Figure 4.1 gives characteristic fouling curves. The deposition of fouling depends on the mechanism of fouling. The removal rate model depends on the entrainment rate, which is proportional to the shear stress at the surface. The fouling deposits also depend on heat exchanger operating parameters like bulk fluid temperature, surface temperatures, combinations of surface material and fluid, and liquid or gas velocity.

The increase in bulk fluid temperature increases chemical reaction rates and crystallisation deposition rates. Chemical reaction rate escalates with the increase in surface temperature; however, reducing surface temperature increases solidification fouling. Sometimes, the surface acts as biocide to biological fouling. Rough surfaces may promote nucleation-based phenomenon. Surface shear stress and mass transfer coefficient affect the scale deposition and the deposition rate is controlled by particulate diffusion. Heat mass transfer analogy may be drawn to explain fouling mechanism (Kim and Webb 1991 and Chamra and Webb 1994b). They have given a correlation between Schmidt number, heat transfer coefficient, Prandtl number and mass transfer coefficient.

More discussion may be obtained from Watkinson (1991), Chamra (1993), Webb and Li (2000), Li and Webb (2000, 2002), Shen et al. (2015), Kim and Webb (1989) and Webb and Chamra (1991).

Marner and Webb (1983) give a survey and bibliography of gas-side fouling. Gas-side fouling may not be significant as far as thermal impedance is concerned. This is so, if the gas-side surface area is much greater than the liquid-side surface area. However, gas pressure drop increases, thereby decreasing the gas flow rate at

Fig. 4.1 Characteristic fouling curves (Webb and Kim 2005)

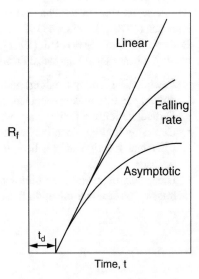

Fig. 4.2 Fouling and cleaning characteristics of the plate-and-fin heat exchangers using 677/149 °C gas inlet/outlet temperatures, with 1600 ppm particle loading (Webb and Kim 2005)

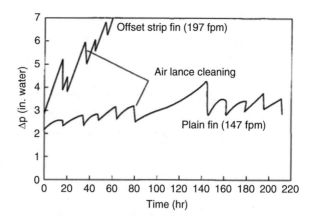

the balance point in the fan curve (Bott and Bemrose 1983). This causes increase of fan power required, which is a costly penalty resulting from the fouling.

Bemrose and Bott (1984) used a multiple regression technique to develop a correlation for the fouling resistance. Industrial heat exchangers used in dusty environment are effected by particulate fouling. Gas-side fouling may not create a serious problem because the ambient air is relatively clear.

Zhang et al. (1990) observed particulate fouling for a plate fin and tube automotive heat exchanger containing oval tubes and louvered fins. Zhang et al. (1990) presented fouling rate curves for different particle loadings. Zhang et al. (1992) observed that turbulence generated by the spoilers at the upstream face reduces fouling. However, the spoilers add to the gas pressure drop.

Webb et al. (1984) studied fouling resulting from combustion products. Burgmeier and Leung (1981) investigated fouling characteristics of a plate and fin geometry using a simulated glass furnace exhaust. Figure 4.2 shows the fouling data for the two surface geometries and fouling and cleaning characteristics of the plate and fin heat exchangers (Burgmeier and Leung 1981). Babuška et al. (2018) worked on $CaCO_3$ deposition on surfaces in heat exchangers and presented a break-off model which considers ageing of the $CaCO_3$ as well as temperature distribution in the layer. The thermal stress causes breaking of fouling materials. This study is very important for industries as $CaCO_3$ depositions are typical phenomenon in heat exchangers.

Roberts and Kubasco (1979) dealt with fouling on a bank of helical finned tubes in the exhaust of a gas turbine. Kindlman and Silvestrini (1979) did not observe measurable soot accumulation on the uncooled tubes. However, the cooled tubes exhibited severe fouling and plugging. If some vapour condenses, a hard deposit on the tube surface forms and it is difficult to remove by air lancing. Of course, this may be water washed. Grillot and Icart (1997) studied diesel exhaust fouling for a circular finned tube heat exchanger. The fouling resistance showed as asymptotic behaviour and the asymptotic fouling resistance increased as the gas velocity decreased and the wall temperature decreased. They have developed the correlations from this data to predict asymptotic fouling resistances and the initial fouling rate.

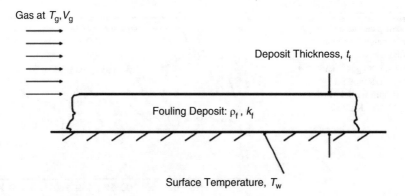

Fig. 4.3 Basic consideration in modelling gas-side fouling (Marner 2014)

Moore (1974) observed that the small axial thermal expansion between adjacent fins acts to break the scale or keep it looser than that on plain tubes. This is based on thin actual examination of fouling deposits and cleaning ability using water jets. Webber (1960) investigated precipitation fouling on integral fin tubes. Finned tubes do not foul as readily as plain tubes and thus are easier to clean; this being so, Moore (1974) and Webber (1960) recommend integral fin tubes for service in dirty environment. Finned tube bundles may be operated for a longer time period before heat transfer limitations warrant necessary cleaning. Katz et al. (1954) and Watkinson (1990) may be referred to for the comparison of performance of plain and finned tube bundles.

Sheikholeslami and Watkinson (1986), Freeman et al. (1990), Müller-Steinhagen et al. (1988) and Gomelauri et al. (1992) have worked with axial fins and ribs in annulus. Owen et al. (1987) investigated particulate fouling effects occurring in gas-cooled nuclear reactors which use fuel rods having transverse rib roughness. Submicron deposits have been observed by them. They have developed a model for prediction of deposition process which accounts for the effect of thermophoresis and surface roughness. Thermophoresis significantly affects the deposition of submicron particles.

Marner (2014) presented predictive overview on gas-side fouling, pressure drop and heat transfer. Heat transfer is reduced due to fouling because of the thermal resistance of the fouling layer. The basic model of gas-side fouling is presented in Fig. 4.3. The effect of gas velocity on particulate fouling in exhaust gas recirculation cooler used in diesel engine application was investigated by Abd-Elhady et al. (2011). The results are shown in Fig. 4.4 in the form of fouling factor R_f versus time with velocity as parameter. Bell et al. (2011) and Ahn and Cho (2003) studied particulate fouling for a hybrid dry air-cooled heat exchanger; they have observed increase in pressure drop up to 50% with 600 g of fouling, and for a finned-tube heat exchanger pressure drop increased up to 45% and decreased in heat transfer up to 14%, respectively. Lankinen et al. (2003) and Bell et al. (2011) investigated the air-side fouling in compact heat exchangers.

Fig. 4.4 Effect of gas velocity on gas-side fouling resistance in an exhaust gas recirculation cooler (Abd-Elhady et al. 2011)

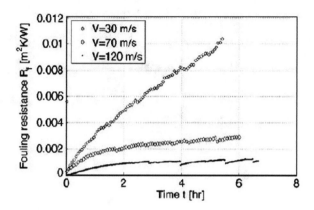

Table 4.1 Geometric details of the dimpled tubes (from Kim 2015)

Tube	D	e	z	P	e/D	z/e	p/e	a	b
Smooth	19.9								
e05z5p3	19.9	0.5	5.0	3.0	0.025	1.0	6.0	2.29	1.70
e05z5p5	19.9	0.5	5.0	5.0	0.025	1.0	10.0	2.29	1.70
e05z5p7	19.9	0.5	5.0	7.0	0.025	14.0	10.0	2.29	1.70
e05z3p3	19.9	0.5	3.0	3.0	0.025	6.0	6.0	2.29	1.70
e05z7p3	19.9	0.5	7.0	3.0	0.025	14.0	6.0	2.29	1.70
e04z5p3	19.9	0.4	5.0	3.0	0.020	12.5	7.5	2.11	1.53
e06z5p3	19.9	0.6	5.0	3.0	0.030	8.3	5.0	2.57	1.93

Kim (2015) investigated the heat transfer performance and fouling characteristics by using three-dimensional dimpled tube with electric steam condenser. He compared the effect of three-dimensional with two-dimensional roughened tube on heat transfer performance and frictional characteristics. He optimized the best configuration of dimple, which maximizes the thermal efficiency.

Table 4.1 shows the geometric details of dimpled tubes. Dimple height of 0.5 mm, axial dimple pitch of 3.0 mm and circumferential dimple pitch of 5.0 mm ($p/e = 6.0$ and $z/e = 10.0$) gave maximum thermal efficiency, which was more than that with the same configuration of tube with two-dimensional spiral rib or the tube with three-dimensional diamond-shaped roughness. Figure 4.5 illustrates that the effect of dimple height on both heat transfer coefficients and friction factors of the dimpled tubes is more than that of smooth tube. Table 4.2 shows the comparison of thermal performance of the specific enhanced tube with other enhanced tubes. Figure 4.6 shows the variation of thermal efficiency with axial dimple pitch.

Takahashi et al. (1988) presented experimental results on the performance of tubes having three-dimensional ribbed geometry. Liao et al. (2000) investigated the tubes having three-dimensional integral roughness with triangular cross section. Local heat transfer measurements of dimpled surfaces were conducted by Mahmood and Ligrani (2002) and Nishida et al. (2012). He et al. (2016) worked with a rotor-assembled strand tube insert to improve heat transfer augmentation and scale

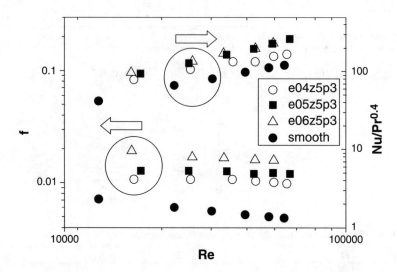

Fig. 4.5 Graph showing the effect of dimple height on the heat transfer coefficient and the friction factor (Kim 2015)

Table 4.2 Thermal performance comparison with other enhanced tubes ($Re = 25{,}000$, $Pr = 10.4$) (from Kim 2015)

Table	D	e/D	p/e	z/e	h/h_p	f/f_p	η
Turbo-BIII	16.4	0.025	3.2	–	2.54	2.30	1.10
A-8	14.0	0.036	7.6	7.6	3.75	3.35	1.11
e05z5p3	19.9	0.025	6.0	10.0	3.03	2.12	1.15

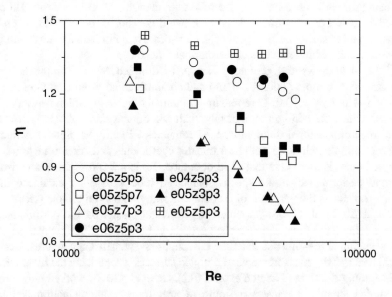

Fig. 4.6 Thermal efficiency of the dimpled tubes (Kim 2015)

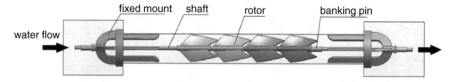

Fig. 4.7 Schematic diagram of the rotor-assembled strand (Yang et al. 2005)

prevention in a heat exchanger. They studied the effect of stress distribution and pressure drop variation on a fixed mounting, which is an important part of the rotor-assembled strand. The rotor-assembled strand shown in Fig. 4.7 was first proposed by Yang et al. (2005).

4.2 Liquid Fouling in Internally Finned Tubes, Rough Tubes, Plate-Fin Geometry and Fouling in Plate Heat Exchanger

Watkinson et al. (1974) and Watkinson and Martinez (1975) investigated liquid fouling on internally finned and spirally indented tubes. Table 4.3 and Fig. 4.8 show asymptotic fouling resistances for internally finned and plain tubes. Somerscales et al. (1991) and Panchal (1989) are additional references from which further information on liquid fouling in internally finned tubes can be obtained.

Webb and Kim (1989), Lietner (1980), Webb and Chamra (1991), Keysselitz (1984), Renfftlen (1991) and Somerscales et al. (1991) experimentally investigated accelerated particulate fouling in rough tubes. Long-term fouling with Wolverine Korodense tube has been investigated by Boyd et al. (1983), Rabas et al. (1990), Li and Webb (2000) and Webb and Li (2000).

Table 4.4 shows that fouling resistance increases as the number of internal ridges inside the tube increases. Figure 4.9 shows the amount of deposit versus Reynolds number for the plate-fin heat exchanger obtained by Masri and Cliffe (1996) in connection with liquid fouling in plate-fin geometry. Plate heat exchangers are found to be in wide use in the process and food industry.

Cooper et al. (1980), Müller-Steinhagen and Meddis (1989), Bansal et al. (1997) and Thonon et al. (1999) have investigated fouling in plate heat exchanger. All of them have observed that the fouling resistance is asymptotic and inversely proportional to the fluid velocity square and proportional to the concentration.

The scale formation is highly non-uniform with more severe fouling in the outlet part of the plate, where the wall temperature is the highest. With increasing flow velocity, both the initial fouling rate and the fouling resistance decrease and this is typical for a reaction-controlled deposition. The corrugation angle influences the particulate fouling in plate heat exchanger. Figure 4.10 shows that high corrugation angle leads to low fouling resistance, since the turbulent intensity becomes stronger with the increase in corrugation angle.

Table 4.3 Ratio of fouling resistances for enhanced and plain tubes (Webb and Kim 2005)

Geometry	n_f	d_i (mm)	e_f/d_i	A_f/L	A/A_B	$R_{f\pm}/R_f$
Plain	0	10.4	0.00	0.033	1.00	1.00
Internal fin 1	6	11.8	0.16	0.052	1.42	1.15
Internal fin 2	10	10.4	0.13	0.060	1.85	1.27
Internal fin 3	16	12.7	0.14	0.082	2.46	1.32
Spirally indented	4	17.4	0.23	0.054	1.0	1.15
Spirally indented	4	17.6	0.24	0.055	1.0	2.30

Fig. 4.8 Asymptotic fouling resistances for internally finned and plain tubes (Watkinson 1990)

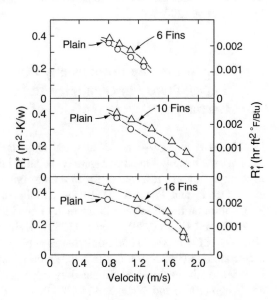

Table 4.4 Fouling resistance vs. tube geometry at the end of 2500-h test period (Webb and Kim 2005)

Tube	$R_f \times 10^4$ (m²-K/W)	$R_f \times 10^4$ (repeat test) (m²-K/W)	n_f	e (mm)	α (°)	p/e	h/h_p
2	1.44	1.71	45	0.33	45	2.81	2.32
5	0.95	1.0	40	0.47	35	3.31	2.26
3	0.63	0.92	30	0.40	45	3.50	2.33
6	0.44	0.46	25	0.49	35	5.02	2.08
7	0.42	0.42	25	0.53	25	7.05	1.93
8	0.35	N/A	18	0.55	25	9.77	1.51
4	0.32	0.49	10	0.43	43	9–88	1.74
1	0.28	0.40	NA	NA	NA	NA	1.0

Fig. 4.9 Variation amount
of deposition with Reynolds
number for plate-fin heat
exchanger (Masri and Cliffe
1996)

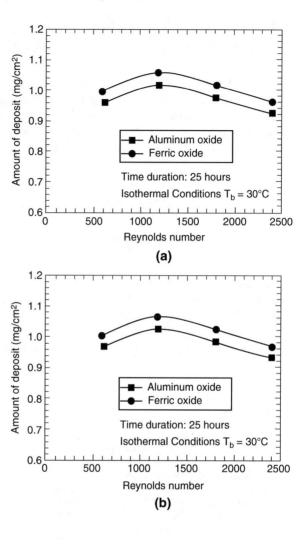

(a)

(b)

Fig. 4.10 Effect of
corrugation angle on CaC0$_3$
particulate fouling at
4.4 kg/m' concentration
in the plate heat exchanger
(Thonon et al. 1999)

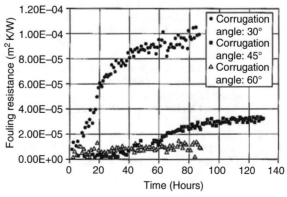

4.3   Modelling of Fouling in Enhanced Tubes and Correlations for Fouling in Rough Tubes

Kim and Webb (1990) and Chamra and Webb (1994a) tested transverse rib-roughened tubes. Kim and Webb (1991), Chamra (1993) and Chamra and Webb (1994b) have developed predictive models for particulate fouling in rough tubes. In the model, the wall shear stress was predicted but the analytical model was proposed by Lewis (1975). Chamra and Webb (1994b) have given the following correlations:

$$\frac{S_{enh}}{S_{ref}} \propto \tau_w^{-0.721} d_p^{-0.319} C_b^{1.02} (e/D)^{-0.307} \qquad (4.2)$$

$$\frac{\xi_{enh}}{\xi_{ref}} \propto \tau_w^{-0.435} d_p^{-0.0769} C_b^{0.421} \left(e/d_i\right)^{-0.396} \qquad (4.3)$$

Further, in the models, the results of plain smooth tubes were obtained by Watkinson and Epstein (1970). However, Kern and Seaton (1959) model underpredicts the experimental data of other investigators with repeated-rib tubes. The difference between the exponents occurs because the results of Chamra and Webb (1994b) are applicable to both the diffusion and inertia regime.

References

Abd-Elhady MS, Zornek T, Malayeri MR, Balestrino S, Szymkowicz PG, Müller-Steinhagen H (2011) Influence of gas velocity on particulate fouling of exhaust gas recirculation coolers. Int J Heat Mass Transfer 54(4):838–846

Ahn YC, Cho JM (2003) An experimental study of the air-side particulate fouling in fin-and tube heat exchangers of air conditioners. J Chem Eng 20:873–877, Korean

Babuška I, Silva RS, Actor J (2018) Break-off model for CaCO3 fouling in heat exchangers. Int J Heat Mass Transfer 116:104–114

Bansal B, Muller-Steinhagen H, Chen XD (1997) Effect of suspended particles on crystallization fouling in plate heat exchangers. J Heat Transfer 119(3):568–574

Bell IH, Groll EA, König H (2011) Experimental analysis of the effects of particulate fouling on heat exchanger heat transfer and air-side pressure drop for a hybrid dry cooler. Heat Transfer Eng 32(3–4):264–271

Bemrose CR, Bott TR (1984) Correlations for gas-side fouling of finned tubes. Institute of Jul 3

Bergles AE, Somerscales EFC (1995) The effect of fouling on enhanced heat transfer equipment. J Enhanc Heat Transf 2:157–166

Bott TR, Bemrose CR (1983) Particulate fouling on the gas-side of finned tube heat exchangers. J Heat Transfer 105(1):178–183

Bott TR (1995) Fouling of heat exchangers. Elsevier

Boyd LW, Hammon JC, Littrel JJ, Withers JG (1983) Efficiency improvement at Gallatin Unit 1 with corrugated condenser tubes. Am Soc Mech Eng 83

Burgmeier L, Leung S (1981) Brayton-cycle heat recovery-system characterization program. Subatmospheric-system test report (No. DOE/CS/40008-T9). Alpha Glass, Inc., El Segundo, CA (USA); AiResearch Mfg. Co., Torrance, CA (USA)

Chamra, LM (1993) A theoretical and experimental study of particulate fouling in enhanced tubes

Chamra LM, Webb RL (1994a) Effect of particle size and size distribution on particulate fouling in enhanced tubes. J Enhanc Heat Transf 1(1):65–75

Chamra LM, Webb RL (1994b) Modeling liquid-side particulate fouling in enhanced tubes. Int J Heat Mass Transfer 37(4):571–579

Cooper A, Suitor JW, Usher JD (1980) Cooling water fouling in plate heat exchangers. Heat Transfer Eng 1(3):50–55

Epstein N (1983a) Thinking about fouling: a 5 × 5 matrix. Heat Transfer Eng 4(1):43–56

Epstein N (1983b) Thinking about heat transfer fouling: a 5 × 5 matrix. Heat Transfer Eng 4 (1):43–56

Epstein N (1988a) General thermal fouling models. In: Fouling science and technology. Springer, Dordrecht, pp 15–30

Epstein N (1988b) Particulate fouling of heat transfer surfaces: mechanisms and models. In: Fouling science and technology. Springer, Dordrecht, pp 143–164

Freeman WB, Middis J, Müller-Steinhagen HM (1990) Influence of augmented surfaces and of surface finish on particulate fouling in double pipe heat exchangers. Chem Eng Process Process Intensif 27(1):1–11

Garrett-Price BA, Smith SA, Watts RL, Knudsen JG, Marner WJ, Suitor JW (1985) Fouling of heat exchangers. Noyes Publications, Park Ridge, NJ

Gomelauri VI, Gruzin AN, Magrakvelidze T, Lekveishvili NN (1992) The effect of two-dimensional artificial roughness on the formation of deposits on heat transfer surfaces. Therm Eng 39(8):439–441

Grillot JM, Icart G (1997) Fouling of a cylindrical probe and a finned tube bundle in a diesel exhaust environment. Exp Therm Fluid Sci 14(4):442–445

He L, Yang W, Guan C, Yan H (2016) Hydrodynamic characteristics and structural improvement of a fixed mount in a heat exchanger with one-way fluid–structure interaction. J Enhanc Heat Transf 23(6):431–447

Katz DL, Knudsen JG, Balekjian G, Grover SS (1954) Fouling of heat exchangers. Petroleum Refiner 33(4):123–125

Kern DQ, Seaton RE (1959) A theoretical analysis of thermal surface fouling. Br Chem Eng 4:258–262

Keysselitz J (1984) Can waterside condenser fouling be controlled operationally. ASME Heat Transfer Div 35:47–54

Kim NH (2015) Single-phase pressure drop and heat transfer measurements of turbulent flow inside helically dimpled tubes. J Enhanc Heat Transf 22(4):345–363

Kim NH, Webb RL (1989) Experimental study of particulate fouling in enhanced water chiller condenser tubes. ASHRAE Transactions (American Society of Heating, Refrigerating and Air-Conditioning Engineers);(USA), 95(CONF-890609)

Kim NH, Webb RL (1990) Particulate fouling inside tubes having arc-shaped two-dimensional roughness by flowing suspension of aluminium oxide in water. Heat Transfer, pp 139–146

Kim NH, Webb RL (1991) Particulate fouling of water in tubes having a two-dimensional roughness geometry. Int J Heat Mass Transf 34:2727–2738

Kindlman L, Silverstrini R (1979) Heat exchanger fouling and corrosion evaluation, Air Research Mfg. Co. report 78–1516(2) on DOE Contract DE-AC03-77ET11296, April 30

Lankinen R, Suihkonen J, Sarkomaa P (2003) The effect of air side fouling on thermal-hydraulic characteristics of a compact heat exchanger. Int J Energy Res 27(4):349–361

Leitner G (1980) Controlling chiller tube fouling. Ashrae J 22(2):40–43

Lewis MJ (1975) An elementary analysis for predicting the momentum-and heat-transfer characteristics of a hydraulically rough surface. J Heat Transfer 97(2):249–254

Li W, Webb RL (2000) Fouling in enhanced tubes using cooling tower water: Part II: Combined particulate and precipitation fouling. Int J Heat Mass Transfer 43(19):3579–3588

Li W, Webb RL (2002) Fouling characteristics of internal helical-rib roughness tubes using low-velocity cooling tower water. Int J Heat Mass Transf 45:1685–1691

Liao Q, Zhu X, Xin MD (2000) Augmentation of turbulent convective heat transfer in tubes with three-dimensional internal extended surfaces. J Enhanc Heat Transf 7(3):139–151

Mahmood GI, Ligrani PM (2002) Heat transfer in a dimpled channel: combined influences of aspect ratio, temperature ratio, Reynolds number, and flow structure. Int J Heat Mass Transfer 45 (10):2011–2020

Marner WJ (2014) Predictive methods for gas-side fouling. J Enhanc Heat Transf 21:4–5

Marner WJ, Webb RL (1983) A bibliography on gas-side fouling. In: Proceedings of the ASME-JSME thermal engineering joint conference 1:1

Masri MA, Cliffe KR (1996) A study of the deposition of fine particles in compact plate fin heat exchangers. J Enhanc Heat Transf 3(4):259–272

Melo LF, Bott TR, Bernardo CA (1987) Fouling science and technology, proceedings of the NATO advanced study institute. Kluwer Academic Publishers, Hingham, MA

Moore JA (1974) Fin tubes foil fouling for scaling services. Chemical Processing (1980)

Müller-Steinhagen HM, Middis J (1989) Particulate fouling in plate heat exchangers. Heat Transfer Eng 10(4):30–36

Müller-Steinhagen H, Reif F, Epstein N, Watkinson AP (1988) Influence of operating conditions on particulate fouling. Can J Chem Eng 66(1):42–50

Nishida S, Murata A, Saito H, Iwamoto K (2012) Compensation of three-dimensional heat conduction inside wall in heat transfer measurement of dimpled surface by using transient technique. J Enhanc Heat Transf 19(4):331–341

Owen I, El-Kady A, Cleaver JW (1987) Fine particle fouling of roughened heat transfer surfaces. In: Proc. 2nd ASME-JSME thermal engineering joint conference, Hawaii, pp 95–101

Panchal CB (1989) Experimental investigation of seawater biofouling for enhanced surfaces (No. CONF-890819-18). Argonne National Lab, Argonne, IL

Renfftlen RG (1991) On-line sponge ball cleaning of enhanced heat transfer tubes. In: National heat transfer conference, HTD 164

Rabas TJ, Merring R, Schaefer R, Lopez-Gomez R, Thors P (1990) Heat-rate improvements obtained with the use of enhanced tubes in surface condensers, presented at the EPRI condenser technology Conf, Boston, 1990

Roberts PB, Kubasco AJ (1979) Combined cycle steam generator gas side fouling evaluation. Phase 1 (No. SR79-R-4557-20). Solar Turbines International, San Diego, CA

Sheikholeslami R, Watkinson AP (1986) Scaling of plain and externally finned heat exchanger tubes. J Heat Transfer 108(1):147–152

Shen C, Cirone C, Jacobi AM, Wang X (2015) Fouling of enhanced tubes for condensers used in cooling tower systems: a literature review. Appl Therm Eng 79:74–87

Somerscales EFC, Ponteduro AF, Bergles AE (1991) Particulate fouling of heat transfer tubes enhanced on their inner surface. In: Fouling and enhancement interactions, vol 164. ASME, New York, pp 17–28

Somerscales EFC, Knudsen JG (1979) Fouling of heat transfer equipment. Hemisphere Publishing Corporation, Washington, DC

Takahashi K, Nakayama W, Kuwahara H (1988) Enhancement of forced convective heat transfer in tubes having three-dimensional spiral ribs. Heat Transfer Jpn Res 17(4):12–28

Thonon B, Grandgeorge S, Jallut C (1999) Effect of geometry and flow conditions on particulate fouling in plate heat exchangers. Heat Transfer Eng 20(3):12–24

Watkinson AP (1990) Fouling of augmented heat transfer tubes. Heat Transfer Eng 11(3):57–65

Watkinson AP (1991) Interactions of enhancement and fouling. In: Fouling and enhancement interactions, vol 164. ASME, New York, pp 1–7

Watkinson AP, Epstein N (1970) In: Proceed. 4th. inter. heat transfer confer., Versailles, France, vol 1, pp 1–12

Watkinson AP, Martinez O (1975) Scaling of spirally indented heat exchanger tubes. J Heat Transfer 97(3):490–492

Watkinson AP, Louis L, Brent R (1974) Scaling of enhanced heat exchanger tubes. Can J Chem Eng 52(5):558–562

Webb RL, Chamra LM (1991) On-line cleaning of particulate fouling in enhanced tubes. In: Fouling and enhancement interactions. ASME, New York

Webb RL, Kim NH (1989) Particulate fouling in enhanced tubes. In: National heat transfer conference

Webb RL, Kim NY (2005) Principles of enhanced heat transfer. Taylor and Francis, New York

Webb RL, Li W (2000) Fouling in enhanced tubes using cooling tower water: Part I: Long-term fouling data. Int J Heat Mass Transfer 43(19):3567–3578

Webb RL, Marchiori D, Durbin RE, Wang YJ, Kulkarni AK (1984) Heat exchangers for secondary heat recovery from glass plants. J Heat Recov Syst 4(2):77–85

Webber WO (1960) Does fouling rule out using finned tubes in reboilers. Petroleum Refiner 39 (3):183–186

Yang WM, Ding YM, Geng LB, Huang W (2005) Rotor-assembled automaticcleaning and heat transfer enhancement device, CN patent 200520127121.9, assigned to Huang Wei and Beijing University of Chemical Technology

Zhang G, Bott TR, Bemrose CR (1990) Finned tube heat exchanger fouling by particles. In: Proc 9th int. heat transfer conf, pp 115–120

Zhang G, Bott TR, Bemrose CR (1992) Reducing particle deposition in air-cooled heat exchangers. Heat Transfer Eng 13(2):81–87

Chapter 5
Conclusions

In this book, introduction to enhanced heat transfer has been discussed in detail. Both passive and active techniques along with their commercial applications have been dealt with. Fundamentals of heat transfer and necessary correction for variation of fluid properties are to be understood clearly to appreciate the impact of heat transfer enhancement on the improved design of compact heat exchangers. This has been followed by complete treatment of fouling on enhanced surface. Fouling fundamentals, fouling of gases on finned surfaces, shell-side fouling of liquids, liquid fouling in internally finned tubes, rough tubes and plate-fin geometry and fouling in plate heat exchanger have been discussed at length. This book ends with the complete treatment of correlations and modelling of fouling in enhanced tubes.

© The Author(s), under exclusive license to Springer Nature Switzerland AG 2020   97
S. K. Saha et al., *Introduction to Enhanced Heat Transfer*, SpringerBriefs in Applied
Sciences and Technology, https://doi.org/10.1007/978-3-030-20740-3_5

Additional References

Carnavos TC (1980) Heat transfer performance of internally finned tubes in turbulent flow. Heat Transfer Eng 1(4):32–37

Chenoweth JM (1990) Final report of the HTRI/TEMA joint committee to review the fouling section of the TEMA standards. Heat Transfer Eng 11(1):73–107

Chiang R (1993) Heat transfer and pressure drop during evaporation and condensation of refrigerant-22 in 7.5 mm and 10 mm diameter axial and helical grooved tubes. In: AIChE symposium series 205-205 American Institute of Chemical Engineers

Eckels SJ, Pate MB (1992) Evaporation heat transfer coefficients for R-22 in micro-fin tubes of different configurations. ASME Papers 177-25

Franke ME, Hogue LE (1991) Electrostatic cooling of a horizontal cylinder. J Heat Transfer 113 (3):544–548

Garrett-Price B (1985) Fouling of heat exchangers: characteristics, costs, prevention, control and removal. Noyes Publications, Park Ridge, p 417

Grassi W, Testi D, Della Vista D (2006) Heat transfer enhancement on the upper surface of a horizontal heated plate in a pool by ion injection from a metallic point. J Electrostat 64:574–580

Ha S, Bergles AE (1993) The influence of oil on local evaporation heat transfer inside a horizontal microfin tube. ASHRAE Trans 99:1244–1255

Ito M, Kimura H (1979) Boiling heat transfer and pressure drop in internal spiral-grooved tubes. Bull JSME 22(171):1251–1257

Jensen MK, Bensler HP (1986) Saturated forced-convective boiling heat transfer with twisted-tape inserts. J Heat Transfer 108(1):93–99

Khanpara JC, Bergles AE, Pate MB (1987) A comparison of in-tube evaporation of refrigerant 113 in electrically heated and fluid heated smooth and inner-fin tubes. Iowa State University College of Engineering

Koyama S, Yu J, Momoki S, Fujii T, Honda H (1995) Forced convective flow boiling heat transfer of pure refrigerants inside a horizontal microfin tube. In: Proceedings of the convective flow boiling, pp 137–142

Kumar A, Layek A (2018) Nusselt number-friction characteristic for a twisted rib roughened rectangular duct using liquid crystal thermography. Exp Therm Fluid Sci 97:205–217

Li HY, Huang RT, Sheu WJ, Wang CC (2007) EHD enhanced heat transfer with needle-arrayed electrodes. In: Proc. of 23th annual IEEE semiconductor and thermal measurement and management symposium, San Jose, CA, pp 149–154

Marner WJ (1990) Progress in gas-side fouling of heat-transfer surfaces. Appl Mech Rev 43 (3):35–66

McCabe WL, Robinson CS (1924) Evaporator scale formation. Ind Eng Chem 16:478–479

© The Author(s), under exclusive license to Springer Nature Switzerland AG 2020
S. K. Saha et al., *Introduction to Enhanced Heat Transfer*, SpringerBriefs in Applied Sciences and Technology, https://doi.org/10.1007/978-3-030-20740-3

Melo L, Bott TR, Bernardo CA (eds) (2012) Fouling science and technology. Springer Science & Business Media, The Netherlands

Metzger D, Fan C, Plevich C (1988) Effects of transverse rib roughness on heat transfer and pressure losses in rectangular ducts with sharp 180 deg turns. In: 26th aerospace sciences meeting, paper no. AIAA-88-0166, Reno, NV

Nae-Hyun K, Webb RL (1991) Particulate fouling of water in tubes having a two-dimensional roughness geometry. Int J Heat Mass Transfer 34(11):2727–2738

Owsenek BL, Seyed-Yagoobi J (1997) Theoretical and experimental study of electrohydrodynamic heat transfer enhancement through wire-plate corona discharge. J Heat Transfer 119(3):604–610

Pascual CC, Stromberger JH, Jeter SM, Abdel-Khalik SI (2000) An empirical correlation for electrohydrodynamic enhancement of natural convection. Int J Heat Mass Transfer 4311:1965–1974

Rabas T, Merring R (1990) Heat-rate improvements obtained with the use of enhanced tubes in surface condensers (No. CONF-9009230-2). Argonne National Lab, Argonne, IL

Schlager LM, Pate MB, Bergles AE (1990) Evaporation and condensation heat transfer and pressure drop in horizontal, 12.7-mm microfin tubes with refrigerant 22. J Heat Transfer 112(4):1041–1047

Singh A, Ohadi MM, Dessiatoun S (1996) Flow boiling heat transfer coefficients of R-134a in a microfin tube. J Heat Transfer 118(2):497–499

Somerscales EFC, Knudsen JG (1981) Fouling of heat transfer equipment. Hemisphere Pub., Washington

Taborek J, Aoki T, Ritter RB, Palen JW, Knudsen JG (1972) Fouling-the major unresolved problem in heat transfer. Chem Eng Prog 68(2, 7):59–67, 69–78

Thors P, Bogart J (1994) In-tube evaporation of HCFC-22 with enhanced tubes. J Enhanc Heat Transf 1(4):365–377

Tubular Exchangers Manufacturers Association (1978) Standards, 6th edn. TEMA, New York, NY

Zimparov V (2002) Enhancement of heat transfer by a combination of a single-start spirally corrugated tubes with a twisted tape. Exp Therm Fluid Sci 25:535–546

Index

© The Author(s), under exclusive license to Springer Nature Switzerland AG 2020
S. K. Saha et al., *Introduction to Enhanced Heat Transfer*, SpringerBriefs in Applied
Sciences and Technology, https://doi.org/10.1007/978-3-030-20740-3

Printed in the United States
By Bookmasters